Name ____________________

Review 1

Add.

1.

2	4	2	0	5
+4	+1	+1	+3	+0

Subtract.

2.

3	4	5	5	6
−2	−3	−1	−2	−3

3. Write a number sentence.

_____ + _____ = _____

Use before pages 9–10.

Name ____________________

Daily Review 2

Count to find how much.

1. 

_____¢

2. 

_____¢

3. 

_____¢

Count on to find how much in all.

4. 

_____¢

5. 

_____¢

6. 

_____¢

Use before pages 11–12.

Name ______

Write the missing numbers.

1.	3,	___,	5, 6,	___,	8
2.	15,	16,	___,	18,	___
3.	53,	___,	55,	___,	57
4.	89,	___,	91,	92,	___

Ring the number that is greater.

5. 4 2

6. 9 12

8. 87 78

9. 64 67

Name

Find the greater number.
Then add by counting on.

1. $\begin{array}{r} 4 \\ +2 \\ \hline \end{array}$ $\begin{array}{r} 2 \\ +4 \\ \hline \end{array}$

2. $\begin{array}{r} 3 \\ +6 \\ \hline \end{array}$ $\begin{array}{r} 6 \\ +3 \\ \hline \end{array}$

3. $\begin{array}{r} 1 \\ +9 \\ \hline \end{array}$ $\begin{array}{r} 9 \\ +1 \\ \hline \end{array}$

4. $\begin{array}{r} 4 \\ +3 \\ \hline \end{array}$ $\begin{array}{r} 3 \\ +4 \\ \hline \end{array}$

Add.

5. $\begin{array}{r} 1 \\ +2 \\ \hline \end{array}$ $\begin{array}{r} 4 \\ +1 \\ \hline \end{array}$ $\begin{array}{r} 6 \\ +2 \\ \hline \end{array}$ $\begin{array}{r} 8 \\ +2 \\ \hline \end{array}$ $\begin{array}{r} 2 \\ +3 \\ \hline \end{array}$

6. $\begin{array}{r} 5 \\ +3 \\ \hline \end{array}$ $\begin{array}{r} 7 \\ +2 \\ \hline \end{array}$ $\begin{array}{r} 3 \\ +1 \\ \hline \end{array}$ $\begin{array}{r} 5 \\ +2 \\ \hline \end{array}$ $\begin{array}{r} 6 \\ +1 \\ \hline \end{array}$

Name ______________________________

Write the sum.

1. $\begin{array}{r} 2 \\ +2 \\ \hline \end{array}$ $\quad \begin{array}{r} 2 \\ +3 \\ \hline \end{array}$ $\quad \begin{array}{r} 3 \\ +2 \\ \hline \end{array}$

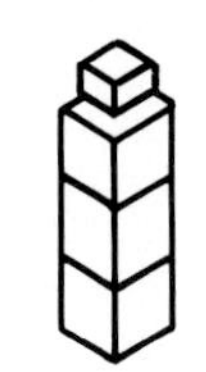

2. $\begin{array}{r} 4 \\ +4 \\ \hline \end{array}$ $\quad \begin{array}{r} 4 \\ +5 \\ \hline \end{array}$ $\quad \begin{array}{r} 5 \\ +4 \\ \hline \end{array}$

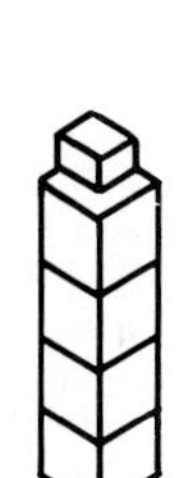
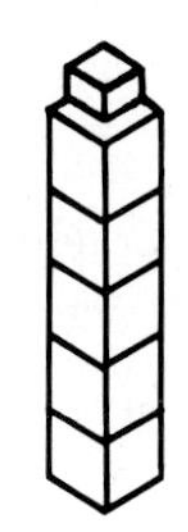

Add by counting on.

3. $\begin{array}{r} 3 \\ +4 \\ \hline \end{array}$ $\quad \begin{array}{r} 1 \\ +2 \\ \hline \end{array}$ $\quad \begin{array}{r} 3 \\ +3 \\ \hline \end{array}$ $\quad \begin{array}{r} 5 \\ +3 \\ \hline \end{array}$ $\quad \begin{array}{r} 8 \\ +2 \\ \hline \end{array}$

4. $\begin{array}{r} 7 \\ +1 \\ \hline \end{array}$ $\quad \begin{array}{r} 4 \\ +2 \\ \hline \end{array}$ $\quad \begin{array}{r} 6 \\ +3 \\ \hline \end{array}$ $\quad \begin{array}{r} 4 \\ +1 \\ \hline \end{array}$ $\quad \begin{array}{r} 7 \\ +2 \\ \hline \end{array}$

Use before pages 17–18.

Name

Add. Ring the sums of 10.

1.	9 + 1	4 + 3	8 + 0	6 + 4	3 + 1
2.	2 + 4	1 + 0	3 + 7	5 + 4	5 + 5
3.	4 + 1	8 + 2	6 + 3	0 + 7	1 + 9

Find the greater number.
Then add by counting on.

4.	8 + 1	2 + 4	3 + 1	2 + 5	3 + 6

Name

Daily Review 7

How many?
Write a number sentence to solve.

1. 3 clowns are smiling.
   2 clowns are frowning.
   How many clowns in all?

_____ + _____ = _____ clowns

2. 4 lions are roaring.
   5 lions are jumping.
   How many lions altogether?

_____ + _____ = _____ lions

Add. Ring the sums of 10.

3.
$$\begin{array}{r} 8 \\ +1 \\ \hline \end{array} \quad \begin{array}{r} 4 \\ +6 \\ \hline \end{array} \quad \begin{array}{r} 3 \\ +2 \\ \hline \end{array} \quad \begin{array}{r} 7 \\ +3 \\ \hline \end{array} \quad \begin{array}{r} 5 \\ +2 \\ \hline \end{array}$$

Name

Add. Ring sums of 10.

1.

4	3	5	2	1
+6	+5	+5	+7	+9

2.

6	7	1	4	8
+0	+3	+9	+5	+2

How many?
Write a number sentence
to solve.

3. 4 kites are soaring.
2 kites are stuck.
How many kites in all?

______ + ______ = ______ kites

Name ____________________

Subtract by counting back.

1.

8	10	5	7	4
−3	−1	−3	−1	−2

2.

7	3	9	5	8
−3	−2	−1	−2	−2

Add. Ring the sums of 10.

3.

3	8	1	5	4
+6	+2	+3	+2	+5

Name

Daily Review 10

Write the difference.

1. $\begin{array}{r} 7 \\ -4 \\ \hline \end{array}$ $\begin{array}{r} 7 \\ -3 \\ \hline \end{array}$

2. $\begin{array}{r} 9 \\ -1 \\ \hline \end{array}$ $\begin{array}{r} 9 \\ -8 \\ \hline \end{array}$

3. $\begin{array}{r} 8 \\ -2 \\ \hline \end{array}$ $\begin{array}{r} 8 \\ -6 \\ \hline \end{array}$

4. $\begin{array}{r} 6 \\ -1 \\ \hline \end{array}$ $\begin{array}{r} 6 \\ -5 \\ \hline \end{array}$

How many?
Write a number sentence
to solve.

5. Sue listens to 4 records.
Alice listens to 2 records.
How many records in all?

_____ + _____ = _____ records

Name

Add or subtract.

1. $\begin{array}{r} 3 \\ +3 \\ \hline \end{array}$ $\begin{array}{r} 6 \\ -3 \\ \hline \end{array}$

2. $\begin{array}{r} 5 \\ +5 \\ \hline \end{array}$ $\begin{array}{r} 10 \\ -5 \\ \hline \end{array}$

Write the difference.

3. $\begin{array}{r} 6 \\ -2 \\ \hline \end{array}$ $\begin{array}{r} 6 \\ -4 \\ \hline \end{array}$

4. $\begin{array}{r} 8 \\ -5 \\ \hline \end{array}$ $\begin{array}{r} 8 \\ -3 \\ \hline \end{array}$

5. $\begin{array}{r} 9 \\ -2 \\ \hline \end{array}$ $\begin{array}{r} 9 \\ -7 \\ \hline \end{array}$

6. $\begin{array}{r} 7 \\ -1 \\ \hline \end{array}$ $\begin{array}{r} 7 \\ -6 \\ \hline \end{array}$

7. $\begin{array}{r} 6 \\ -4 \\ \hline \end{array}$ $\begin{array}{r} 6 \\ -2 \\ \hline \end{array}$

8. $\begin{array}{r} 8 \\ -3 \\ \hline \end{array}$ $\begin{array}{r} 8 \\ -5 \\ \hline \end{array}$

Use before pages 31–32.

Name

Subtract.

1.

$$\begin{array}{r}10\\-2\\\hline\end{array}\quad\begin{array}{r}10\\-4\\\hline\end{array}\quad\begin{array}{r}10\\-9\\\hline\end{array}\quad\begin{array}{r}10\\-5\\\hline\end{array}\quad\begin{array}{r}10\\-7\\\hline\end{array}$$

Add or subtract.

2.

$$\begin{array}{r}4\\+4\\\hline\end{array}\quad\begin{array}{r}8\\-4\\\hline\end{array}$$

3.

$$\begin{array}{r}1\\+1\\\hline\end{array}\quad\begin{array}{r}2\\-1\\\hline\end{array}$$

4.

$$\begin{array}{r}5\\+5\\\hline\end{array}\quad\begin{array}{r}10\\-5\\\hline\end{array}$$

5.

$$\begin{array}{r}3\\+3\\\hline\end{array}\quad\begin{array}{r}6\\-3\\\hline\end{array}$$

6.

$$\begin{array}{r}8\\-4\\\hline\end{array}\quad\begin{array}{r}4\\-2\\\hline\end{array}\quad\begin{array}{r}10\\-5\\\hline\end{array}\quad\begin{array}{r}6\\-3\\\hline\end{array}\quad\begin{array}{r}2\\-1\\\hline\end{array}$$

Name ____________________

Daily Review 13

Add or subtract.

1. $4 + 1 =$ ____

$1 + 4 =$ ____

$5 - 4 =$ ____

$5 - 1 =$ ____

2. $7 + 2 =$ ____

$2 + 7 =$ ____

$9 - 7 =$ ____

$9 - 2 =$ ____

Subtract.

3. $\begin{array}{r} 10 \\ -2 \\ \hline \end{array}$ $\begin{array}{r} 10 \\ -8 \\ \hline \end{array}$

4. $\begin{array}{r} 5 \\ -1 \\ \hline \end{array}$ $\begin{array}{r} 5 \\ -4 \\ \hline \end{array}$

5. $\begin{array}{r} 9 \\ -3 \\ \hline \end{array}$ $\begin{array}{r} 8 \\ -1 \\ \hline \end{array}$ $\begin{array}{r} 5 \\ -2 \\ \hline \end{array}$ $\begin{array}{r} 6 \\ -4 \\ \hline \end{array}$ $\begin{array}{r} 8 \\ -1 \\ \hline \end{array}$

6. $\begin{array}{r} 10 \\ -6 \\ \hline \end{array}$ $\begin{array}{r} 3 \\ -2 \\ \hline \end{array}$ $\begin{array}{r} 9 \\ -4 \\ \hline \end{array}$ $\begin{array}{r} 10 \\ -2 \\ \hline \end{array}$ $\begin{array}{r} 10 \\ -1 \\ \hline \end{array}$

Name

Ring the facts for the number in each balloon.

1. 

$6 + 1$

$7 - 5$

$10 - 3$

2. 

$2 + 2$

$8 - 6$

$1 + 1$

Subtract.

3. $\begin{array}{r} 10 \\ -1 \\ \hline \end{array}$ $\begin{array}{r} 10 \\ -7 \\ \hline \end{array}$ $\begin{array}{r} 10 \\ -5 \\ \hline \end{array}$ $\begin{array}{r} 10 \\ -3 \\ \hline \end{array}$ $\begin{array}{r} 10 \\ -2 \\ \hline \end{array}$

Add and subtract.

4. $\begin{array}{r} 2 \\ +2 \\ \hline \end{array}$ $\begin{array}{r} 4 \\ -2 \\ \hline \end{array}$

5. $\begin{array}{r} 4 \\ +4 \\ \hline \end{array}$ $\begin{array}{r} 8 \\ -4 \\ \hline \end{array}$

Name ______________________

Ring the correct number sentence.

1. Harry found 6 sea shells.
   He gave 4 to a friend.
   How many sea shells does he have left?

   $6 + 4 = 10$ $\quad$ $6 - 4 = 2$

Add or subtract.

2. $2 + 8 =$ ____

   $8 + 2 =$ ____

   $10 - 8 =$ ____

   $10 - 2 =$ ____

3. $2 + 5 =$ ____

   $5 + 2 =$ ____

   $7 - 2 =$ ____

   $7 - 5 =$ ____

Subtract.

4. $\begin{array}{r} 10 \\ -3 \\ \hline \end{array}$ $\quad$ $\begin{array}{r} 8 \\ -6 \\ \hline \end{array}$ $\quad$ $\begin{array}{r} 10 \\ -9 \\ \hline \end{array}$ $\quad$ $\begin{array}{r} 5 \\ -1 \\ \hline \end{array}$ $\quad$ $\begin{array}{r} 9 \\ -3 \\ \hline \end{array}$

Name

Think of the greater number.
Then add by counting on.

1.

9	7	8	6	3
+3	+2	+3	+2	+4

Ring the correct number sentence.

2. Rob picked 7 apples.
Then he picked 2 more.
How many apples does he have in all?

$7 + 2 = 9$ $7 - 2 = 5$

Ring the facts for the number in each star.

3.

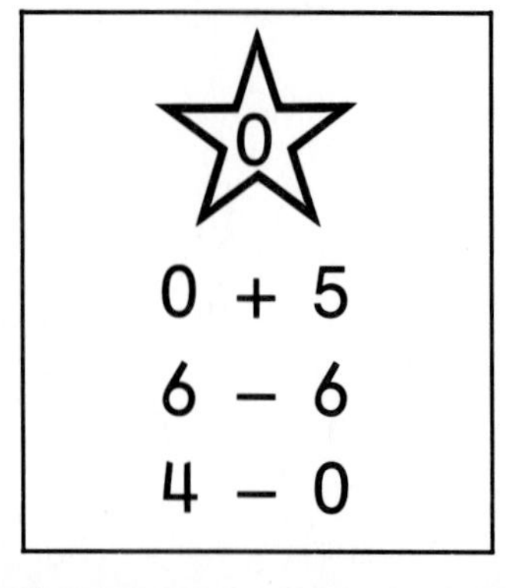

4.

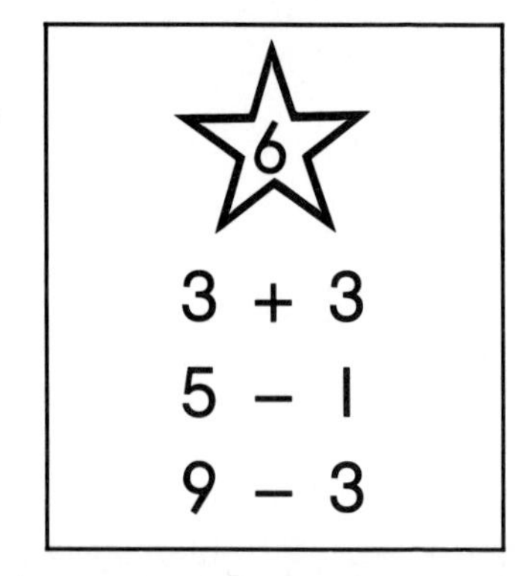

Name

Daily Review 17

Add.

1.

0	7	8	9	2
+0	+7	+4	+9	+7

2.

8	4	6	2	7
+8	+3	+5	+2	+6

Ring the correct number sentence.

3. Jerry ate 5 grapes.
Then he ate 2 grapes.
How many grapes did he eat in all?

$5 + 2 = 7$ $5 - 2 = 3$

Ring the facts for the number in each shape.

4.

9

3 + 6

8 − 1

9 − 0

5.

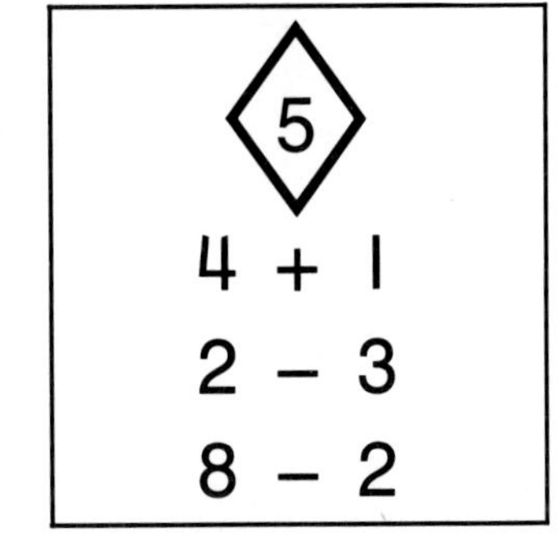

Name

Daily Review 18

Write the sum.

1. $\begin{array}{r} 4 \\ +4 \\ \hline \end{array}$ $\begin{array}{r} 4 \\ +5 \\ \hline \end{array}$ $\begin{array}{r} 5 \\ +4 \\ \hline \end{array}$

2. $\begin{array}{r} 6 \\ +6 \\ \hline \end{array}$ $\begin{array}{r} 6 \\ +7 \\ \hline \end{array}$ $\begin{array}{r} 7 \\ +6 \\ \hline \end{array}$

Find the greater number.
Then add by counting on.

3. $\begin{array}{r} 8 \\ +2 \\ \hline \end{array}$ $\begin{array}{r} 2 \\ +8 \\ \hline \end{array}$

4. $\begin{array}{r} 2 \\ +6 \\ \hline \end{array}$ $\begin{array}{r} 6 \\ +2 \\ \hline \end{array}$

5. $\begin{array}{r} 7 \\ +2 \\ \hline \end{array}$ $\begin{array}{r} 2 \\ +3 \\ \hline \end{array}$ $\begin{array}{r} 9 \\ +2 \\ \hline \end{array}$ $\begin{array}{r} 3 \\ +4 \\ \hline \end{array}$ $\begin{array}{r} 3 \\ +8 \\ \hline \end{array}$

Use before pages 53–54.

Name

1. Make a graph.

Color 1 ☐ for each kind of cat.

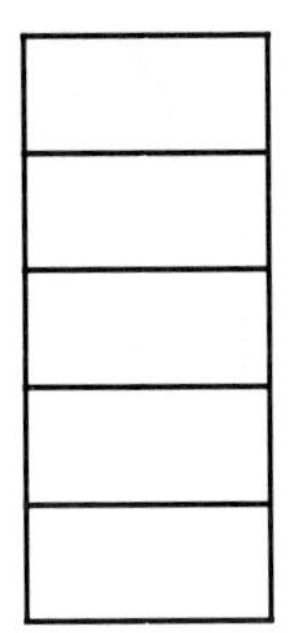 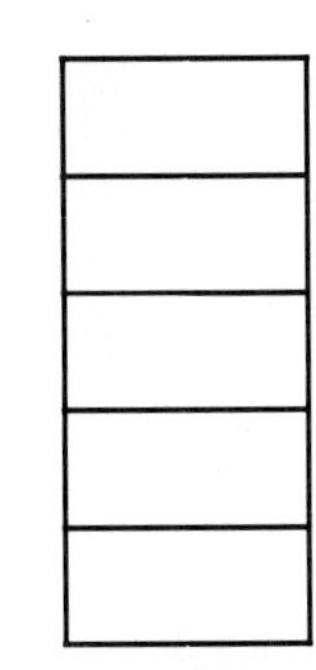 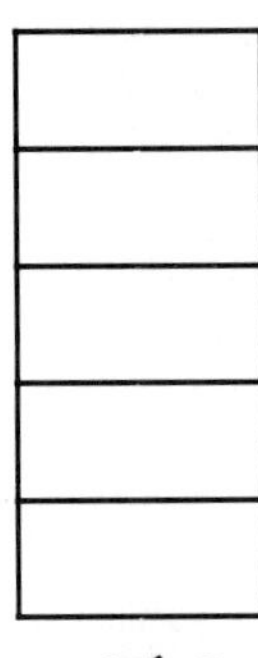

Add.

2.

8	7	5	4	6
+8	+7	+5	+4	+6

Name

Add.

1.

5	9	4	5	8
3	0	2	1	0
+1	+3	+1	+5	+5

2.

0	3	3	7	9
7	6	2	4	5
+3	+2	+1	+4	+0

Write the sum.

3.

5	5	6
+5	+6	+5

4.

3	3	4
+3	+4	+3

5.

7	7	8
+7	+8	+7

6.

4	4	5
+4	+5	+4

Name

Daily Review 21

Add.

1.

2	3	4	7	0
4	4	5	6	9
+3	+5	+1	+0	+3

2.

4	9	8	2	2
6	3	0	7	1
+4	+6	+7	+3	+5

Write the sum.

3.

Add 4	
7	
0	
6	

4.

Add 7	
8	
5	
9	

5.

Add 8	
3	
6	
8	

Name

Add.

1.	9 +7	5 +9	9 +2	6 +9	8 +9
2.	9 +3	9 +5	7 +6	9 +8	8 +8
3.	4 +9	7 +4	6 +7	8 +7	8 +4
4.	0 0 +0	7 4 +7	3 4 +1	5 2 +4	3 5 +5

Name

Daily Review 23

1. Find the sums.
Color 15 red.
Color 16 blue.
Color 17 purple.
Color 18 yellow.

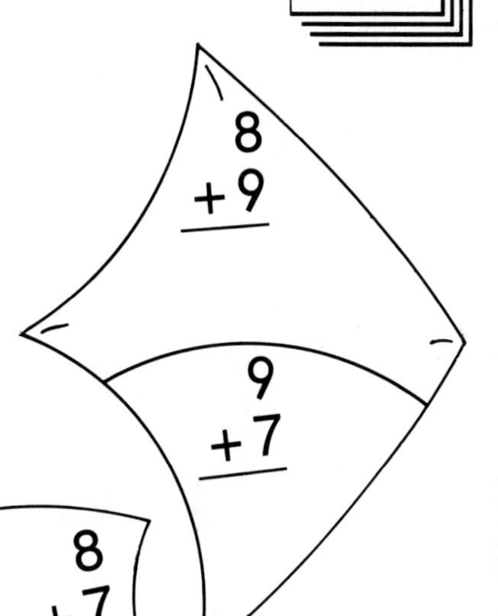

$\begin{array}{r} 9 \\ +9 \\ \hline \end{array}$ $\begin{array}{r} 9 \\ +8 \\ \hline \end{array}$ $\begin{array}{r} 8 \\ +7 \\ \hline \end{array}$

$\begin{array}{r} 9 \\ +6 \\ \hline \end{array}$ $\begin{array}{r} 8 \\ +8 \\ \hline \end{array}$ $\begin{array}{r} 7 \\ +9 \\ \hline \end{array}$

Add.

2. $\begin{array}{r} 5 \\ 4 \\ +5 \\ \hline \end{array}$ $\begin{array}{r} 8 \\ 0 \\ +4 \\ \hline \end{array}$ $\begin{array}{r} 3 \\ 1 \\ +5 \\ \hline \end{array}$ $\begin{array}{r} 0 \\ 9 \\ +0 \\ \hline \end{array}$ $\begin{array}{r} 2 \\ 8 \\ +6 \\ \hline \end{array}$

Name

George and his friends went fishing. This graph shows how many fish each child caught.

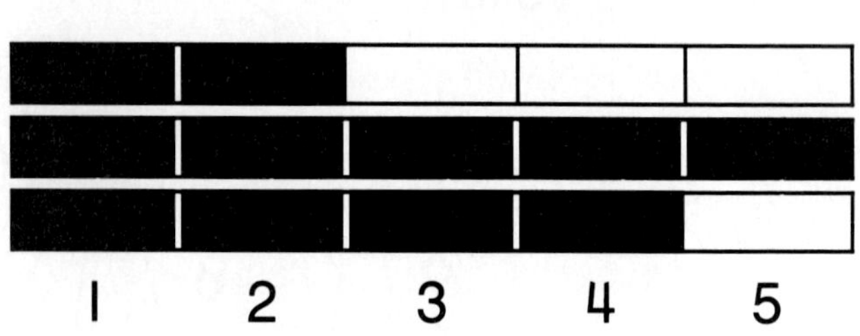

Solve.

1. Who caught the most fish?

2. Who caught 4 fish?

Add.

3.

5	8	7	8	7
+8	+7	+5	+6	+9

Name

Daily Review 25

Books Read in One Week

Jan
Sheila
Chris

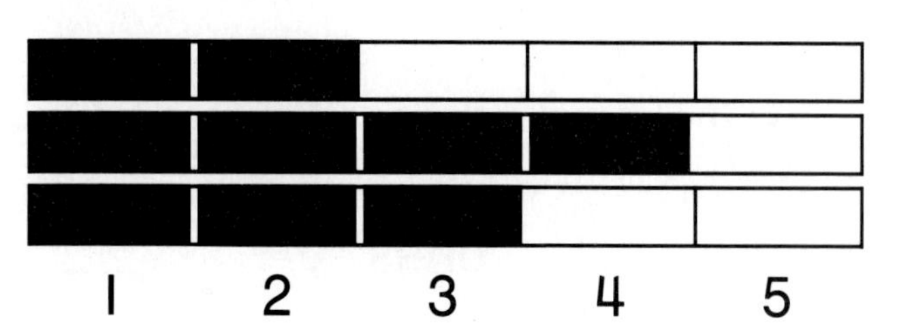

1. Who read the least number of books?

Find the sums.

3. Color 11 green.
4. Color 12 red.
5. Color 13 blue.

Name

How many?

1. 27 ____ tens ____ ones

84 ____ ones ____ tens

Pablo and his friends grew some flowers. This graph shows how many flowers each child grew.

Number of Flowers Grown

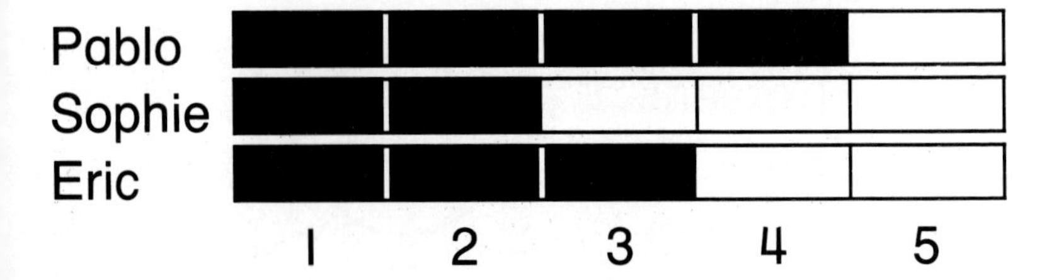

Solve.

2. Who grew the most flowers?

____

3. Who grew two flowers?

____

Name

Write the missing numbers.

1. After	2. Before	3. Between
47 ___	___ 6	25 ___ 27
74 ___	___ 30	60 ___ 62
89 ___	___ 83	89 ___ 91

How many?

4.

___ ones

___ tens

5.

___ tens

___ ones

Use before pages 81–82.

Name

Daily Review 28

Ring the number that is less.

1. 36 69
2. 48 18
3. 80 78
4. 38 25

Write the missing numbers.

5. After	6. Before	7. Between
30 ____	____ 58	59 ____ 61
27 ____	____ 60	82 ____ 84
39 ____	____ 91	31 ____ 33
84 ____	____ 20	68 ____ 70

How many?

8. 59 ____ tens ____ ones

9. 90 ____ tens ____ ones

Name

Daily Review 29

Write > or < in the .

1. 65 ◯ 56 29 ◯ 90 14 ◯ 4
2. 55 ◯ 77 99 ◯ 9 28 ◯ 49
3. 48 ◯ 92 27 ◯ 16 59 ◯ 36

Write the missing numbers.

4. After	5. Before	6. Between
39 ___	___ 78	56 ___ 58
71 ___	___ 74	49 ___ 51
56 ___	___ 88	30 ___ 32
98 ___	___ 30	66 ___ 68
17 ___	___ 83	9 ___ 11

Name ______________________________

What is the pattern?

Ring more numbers to complete the pattern

1.

(0)	1	(2)	3	(4)	5	(6)	7	(8)	9
(10)	11	12	13	14	15	16	17	18	19
20	21	22	23	24	25	26	27	28	29
30	31	32	33	34	35	36	37	38	39
40	41	42	43	44	45	46	47	48	49
50	51	52	53	54	55	56	57	58	59
60	61	62	63	64	65	66	67	68	69

Write > or < in the ◯.

2. 24 ◯ 26　　48 ◯ 84　　88 ◯ 44

3. 8 ◯ 18　　37 ◯ 7　　21 ◯ 11

Name ____________________

Color the elephants.

1. blue: third, sixth, ninth, fifteenth
2. red: second, seventh, thirteenth, fourteenth
3. green: first, eleventh, sixteenth, nineteenth
4. yellow: fourth, eighth, twelfth, eighteenth
5. purple: fifth, tenth, seventeenth, twentieth

Write > or < in the ◯.

6.

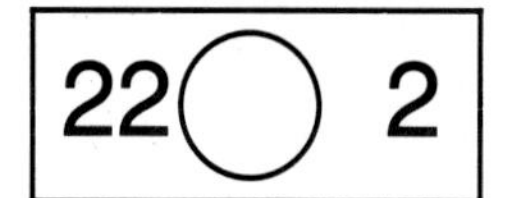

17 ◯ 27

Name

How much in all?

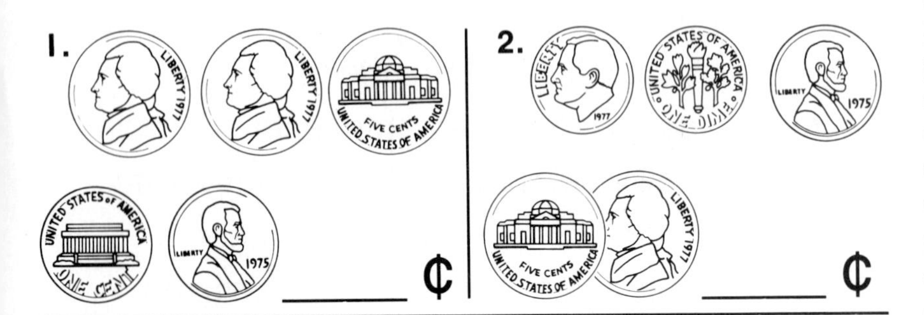

1. ____ ¢

2. ____ ¢

What is the pattern?
Ring more numbers to complete the pattern

3.

(0)	1	2	3	(4)	5	6	7	(8)	9
10	11	(12)	13	14	15	(16)	17	18	19
20	21	22	23	24	25	26	27	28	29
30	31	32	33	34	35	36	37	38	39
40	41	42	43	44	45	46	47	48	49
50	51	52	53	54	55	56	57	58	59
60	61	62	63	64	65	66	67	68	69

Name

How much in all?

1.

______¢

Use crayons to put an X on each jogger.

2. blue: sixth, ninth, thirteenth, eighteenth
3. green: first, eighth, fourteenth, sixteenth
4. yellow: second, seventh, twelfth, nineteenth
5. red: third, fourth, tenth, seventeenth
6. purple: fifth, eleventh, fifteenth, twentieth

Name

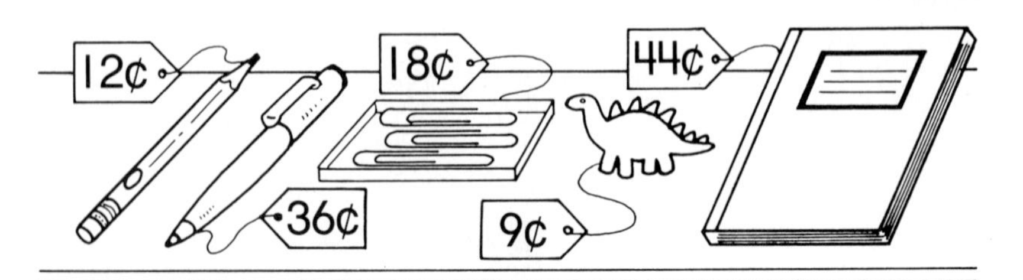

Use the picture of the store.
Ring the coins you need.

1.

2.

How much?

3.

_____ ¢

Name

Daily Review 35

Subtract.

1.

12	18	10	14	16
−6	−9	−5	−7	−8

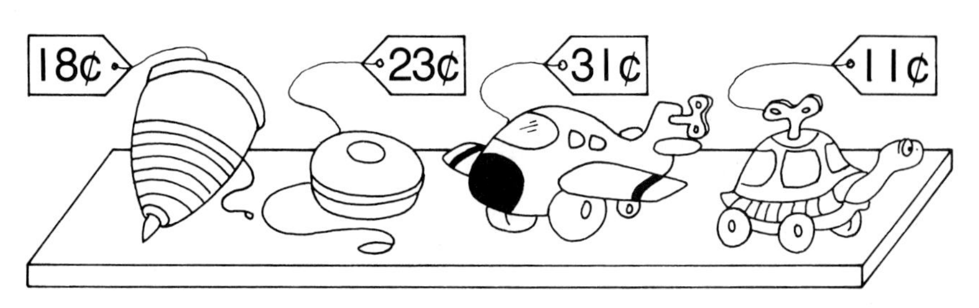

Use the picture of the store.
Ring the coins you need.

2.

3.

Name

Daily Review 36

Show each step. Then solve.

1. 13 tunas are in the water. 4 tunas swim away. How many tunas are left? 3 more tunas swim away. How many are there now?

Step 1

☐ ____

____

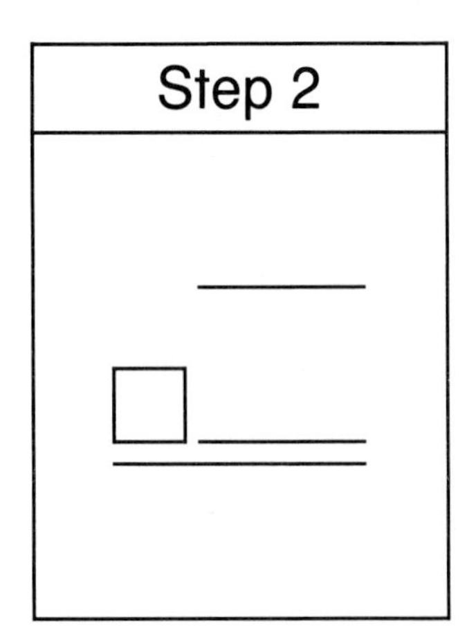

How much in all?

2.

____ ¢

3.

____ ¢

Name

Daily Review 37

Subtract.

1.

11	12	12	11	12
−6	−9	−3	−9	−5

Use the picture.
Ring the coins you need.

2.

3.

Name

Subtract.

1.

14	13	14	13	13
−8	−9	−6	−8	−5

Show each step.
Then solve.

2. 13 boats are sailing in a race. 4 boats tip over. How many boats are still sailing?

2 boats come back in the race. How many are sailing now?

Step 1	Step 2
□ ____	□ ____

Name

Write the difference.

1.

17	15	18	17	16
−8	−9	−9	−9	−8

Show each step. Then solve.

2. Sam had 9 pennies. He earned 5 more pennies. How many pennies did he have altogether?

Sam spent 7 of his pennies. How many pennies does he have left?

Step 1
☐

Step 2
☐

Name

Add or subtract.

1. $7 + 8 =$ ____

$8 + 7 =$ ____

$15 - 8 =$ ____

$15 - 7 =$ ____

2. $5 + 9 =$ ____

$9 + 5 =$ ____

$14 - 5 =$ ____

$14 - 9 =$ ____

Subtract.

3. $\begin{array}{r} 13 \\ -9 \\ \hline \end{array}$ $\begin{array}{r} 14 \\ -8 \\ \hline \end{array}$ $\begin{array}{r} 13 \\ -4 \\ \hline \end{array}$ $\begin{array}{r} 13 \\ -6 \\ \hline \end{array}$ $\begin{array}{r} 14 \\ -9 \\ \hline \end{array}$

4. $\begin{array}{r} 14 \\ -6 \\ \hline \end{array}$ $\begin{array}{r} 13 \\ -7 \\ \hline \end{array}$ $\begin{array}{r} 14 \\ -5 \\ \hline \end{array}$ $\begin{array}{r} 14 \\ -8 \\ \hline \end{array}$ $\begin{array}{r} 14 \\ -7 \\ \hline \end{array}$

5. $\begin{array}{r} 11 \\ -8 \\ \hline \end{array}$ $\begin{array}{r} 12 \\ -5 \\ \hline \end{array}$ $\begin{array}{r} 12 \\ -6 \\ \hline \end{array}$ $\begin{array}{r} 11 \\ -6 \\ \hline \end{array}$ $\begin{array}{r} 12 \\ -8 \\ \hline \end{array}$

Name ______________________

Add or subtract.

1.

Add 6	
5	
3	
9	
6	

2.

Subtract 7	
13	
9	
14	
16	

3.

Add 8	
8	
5	
3	
9	

Write the difference.

4. $\begin{array}{r} 17 \\ -8 \\ \hline \end{array}$ $\begin{array}{r} 15 \\ -9 \\ \hline \end{array}$ $\begin{array}{r} 16 \\ -8 \\ \hline \end{array}$ $\begin{array}{r} 18 \\ -9 \\ \hline \end{array}$ $\begin{array}{r} 13 \\ -8 \\ \hline \end{array}$

5. $\begin{array}{r} 11 \\ -6 \\ \hline \end{array}$ $\begin{array}{r} 12 \\ -9 \\ \hline \end{array}$ $\begin{array}{r} 14 \\ -7 \\ \hline \end{array}$ $\begin{array}{r} 13 \\ -8 \\ \hline \end{array}$ $\begin{array}{r} 12 \\ -6 \\ \hline \end{array}$

Name

Decide if you need to add or subtract. Then solve.

1. Jane reads 7 books. Allie reads 2 books less than Jane. How many books does Allie read?

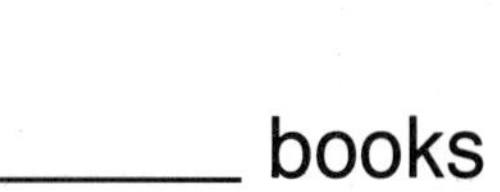

______ books

2. Ron picks 9 flowers. Then he picks 5 more flowers. How many flowers does Ron pick in all?

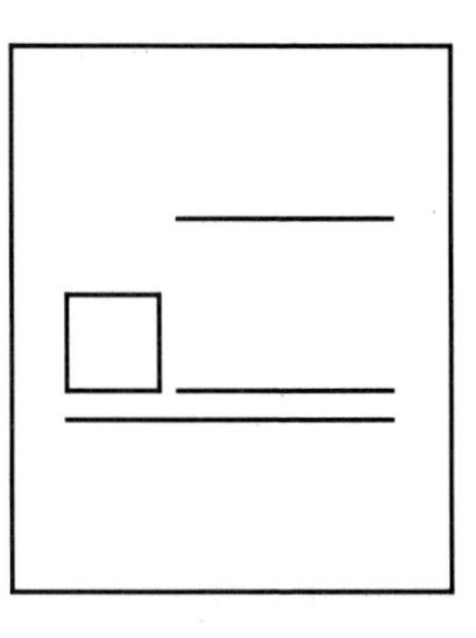

______ flowers

Add or subtract.

3. $5 + 7 =$ ____

$7 + 5 =$ ____

4. $12 - 5 =$ ____

$12 - 7 =$ ____

Name

Daily Review 43

Draw the hands. Write the time.

1.
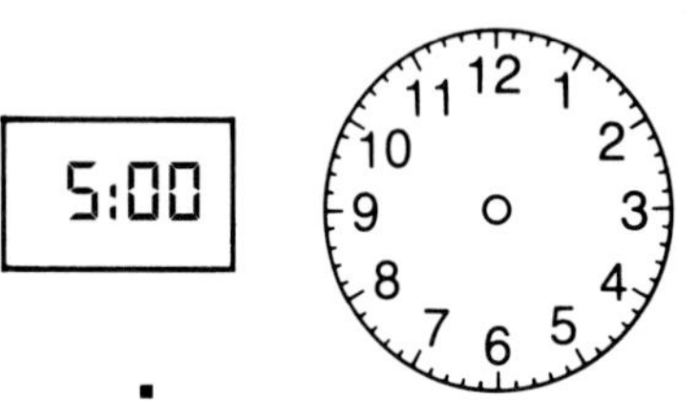

___ : ___

2.
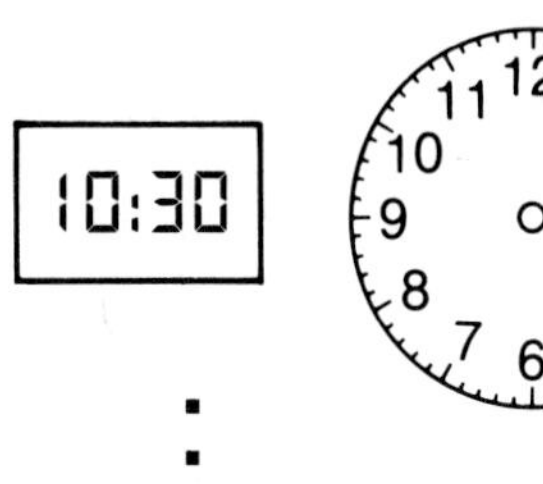

___ : ___

Decide if you need to add or subtract.
Then solve.

3. Sam stacks 15 blocks. Joan stacks 6 blocks less than Sam. How many blocks does Joan stack?

_____ blocks

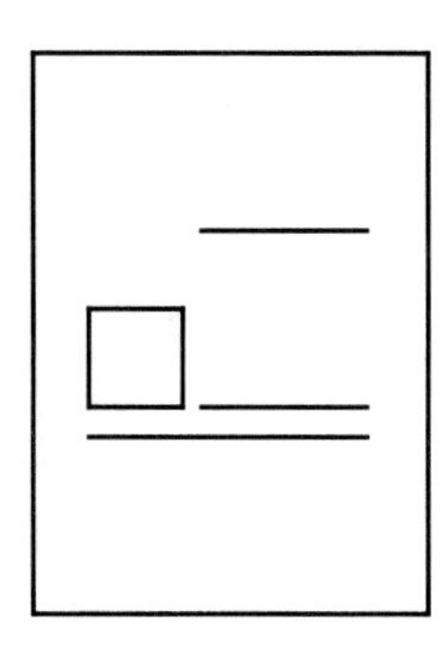

Add or subtract.

4.

Add 5	
6	
9	

5.

Add 9	
4	
9	

6.

Subtract 7	
16	
14	

Name

Count by fives to find the number of minutes.

1. From the 12 to the 4 is

_____ minutes after the hour.

Draw the hands. Write the time.

2.

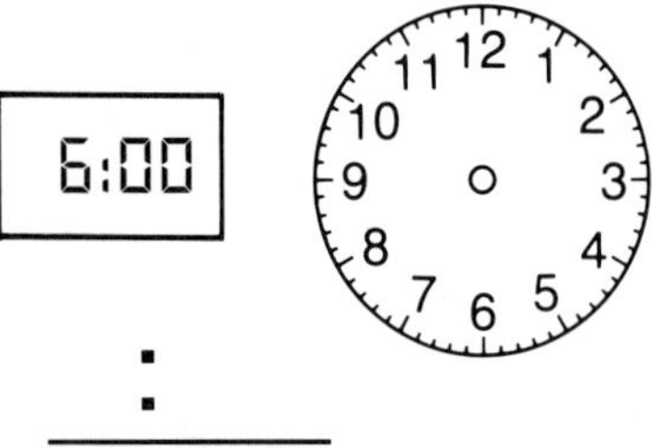

___:___

3.

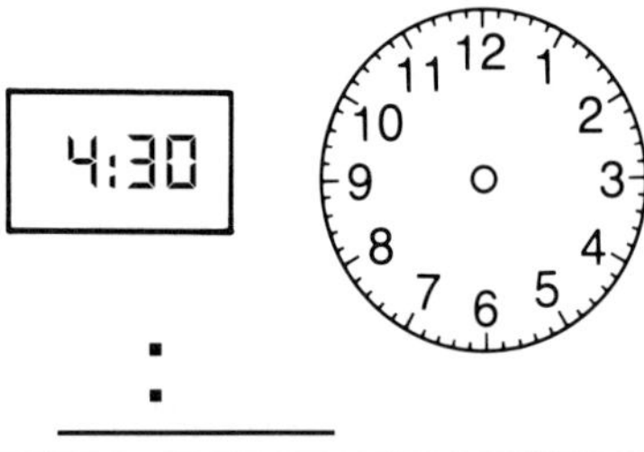

___:___

Add or subtract.

4.

Subtract 6	
15	
12	
14	

5.

Add 5	
6	
8	
9	

6.

Add 8	
9	
7	
3	

Use before pages 137–138.

Name

Daily Review 45

Complete the sentences.
Write the time.

1. 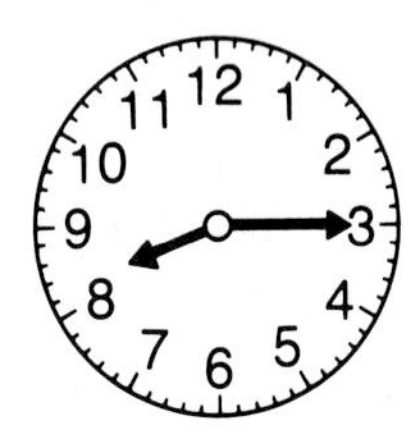

It is ______ minutes

after ______.

It is ___ : ___.

2. 

It is ______ minutes

after ______.

It is ___ : ___.

Count by fives to find
the number of minutes.

3. From the 12 to the 7 is

_____ minutes after the hour.

4. From the 12 to the 10 is

_____ minutes after the hour.

Name ______________________

Daily Review 46

Write the time under each clock.

1. 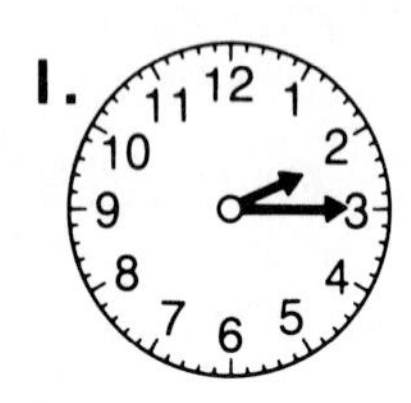

___:___ ___:___ ___:___ ___:___

Count by fives to find the number of minutes.

2. From the 12 to the 8 is

______ minutes after the hour.

3. From the 12 to the 2 is

______ minutes after the hour.

4. From the 12 to the 11 is

______ minutes after the hour.

Name ______________________

House Jobs	Time to complete
Put toys away	30 minutes
Take out the trash	10 minutes

Write the time each job ends.
Use the chart to solve.

1. Put toys away.
Starting time.

Ending time ____:____

2. Take out trash.
Starting time.

Ending time ____:____

Complete the sentence.
Write the time.

3.

It is ______ minutes

after _____ .

It is ____:____.

Name

1991

January	February	March	April
May	June	July	August
September	October	November	December

Use the calendar to answer the questions.

1. Which month is your favorite month? ______

2. Which month comes before your favorite month? ______

3. Which month comes after your favorite month? ______

Write the time next to each clock.

4.

______

______

Use before pages 147–148.

Name

Make a calendar for this month.
Include the days of the week.
Then answer the questions.

Sunday	Monday	Tuesday	Wednesday	Thursday	Friday	Saturday

1. How many days are in this month? ______

2. Ring today's date on your calendar.

3. Today is what day of the week? ______

Name

Use an inch ruler.

About how tall is each object?

First, guess. Then measure to check.

1.

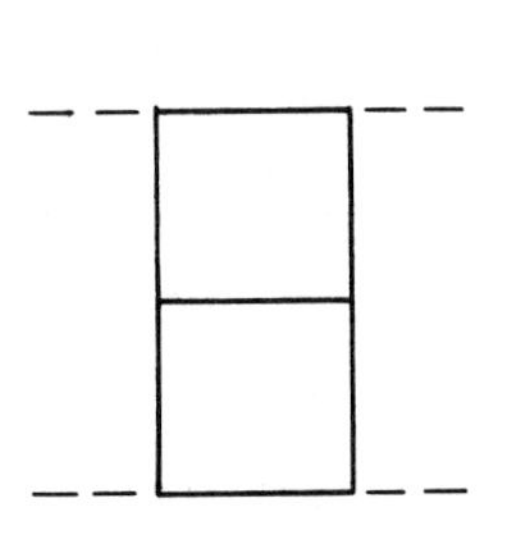

Guess: ______ inches

Measure: ______ inch

2.

Guess: ______ inches

Measure: ____ inches

Write the time under each clock.

3.

____

4.

____

Name

Use an inch ruler.
Find the length to the closer inch.

1.

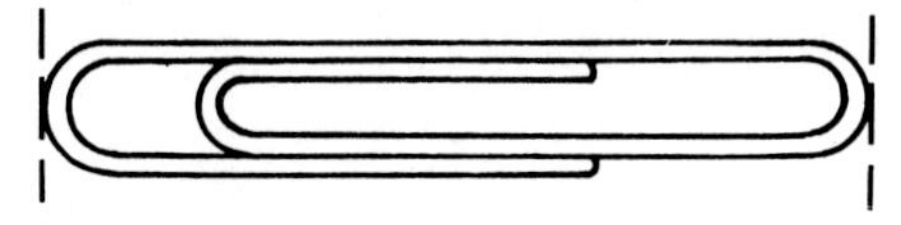

The paper clip is closer to ____ inches.

Use an inch ruler.
First, guess. Then measure to check.

2.

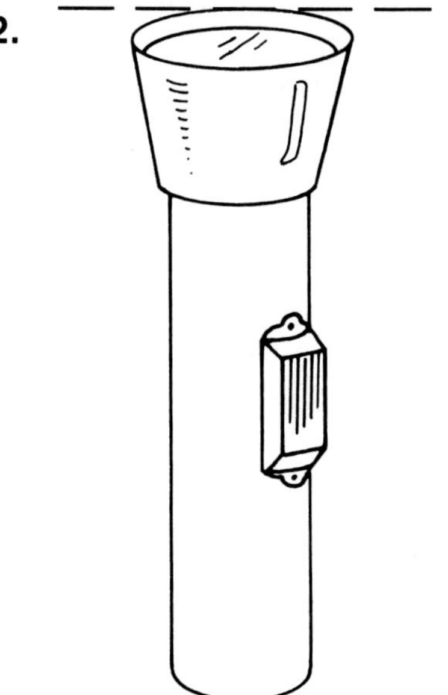

Guess: ______ inches

Measure: ____ inches

3.

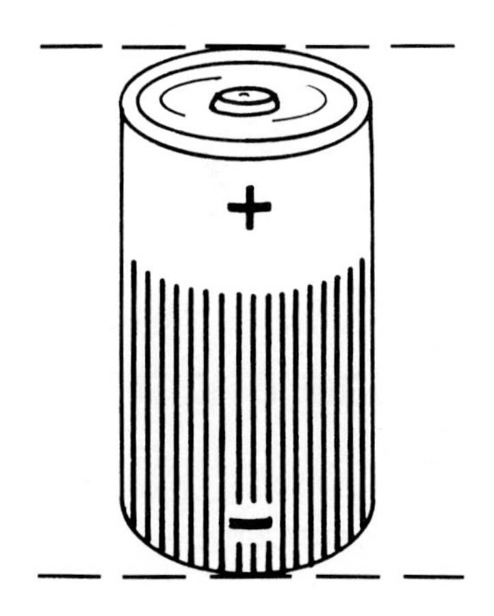

Guess: ______ inches

Measure: ____ inches

Use before pages 153–154.

Name ____________________

1. Ring the objects that are longer than 1 foot.

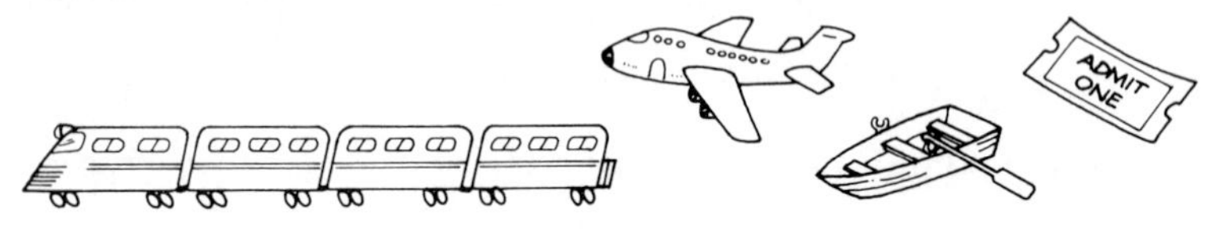

2. Ring the objects that are taller than 1 foot.

3. Ring the objects that are shorter than 1 foot.

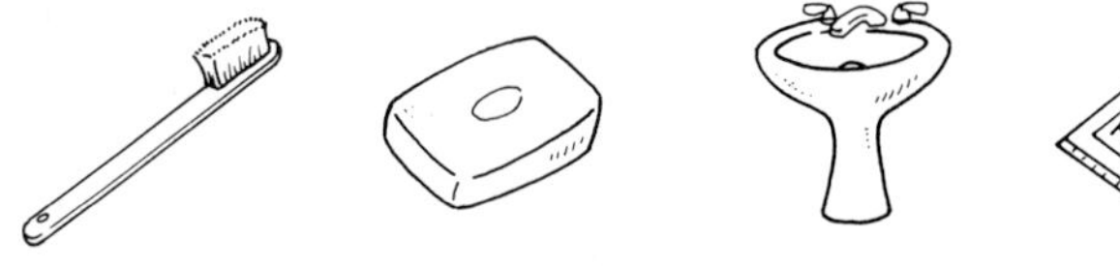

Use an inch ruler.
Find the length to the closer inch.

4.

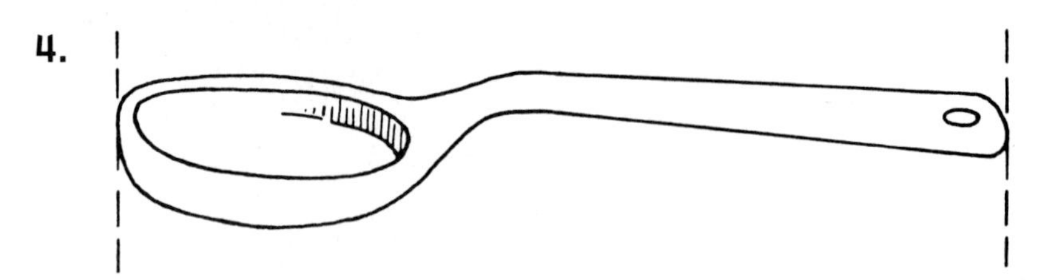

The spoon is closer to _____ inches.

Name

Use a centimeter ruler.
Guess the length of each object.
Then measure to check.

1.

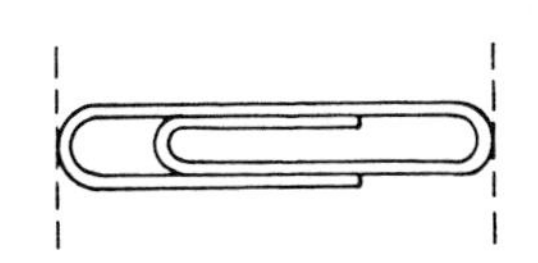

Guess: ________ cm

Measure: ______ cm

2.

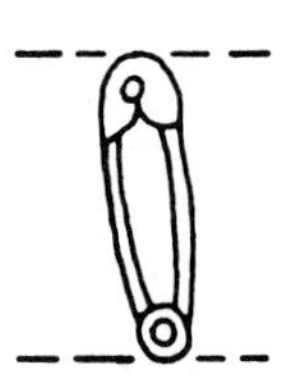

Guess: ________ cm

Measure: ________ cm

Ring the objects that are longer than 1 foot.

3.

Name

Use a centimeter ruler.
Find the length to the closer centimeter.

1.

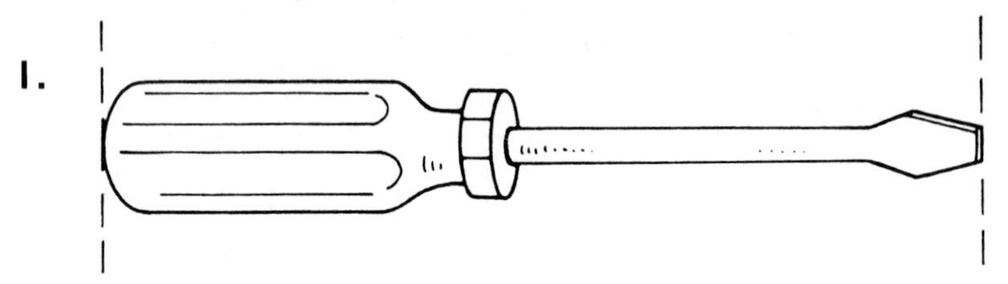

The screwdriver is closer to _____ cm.

Use a centimeter ruler.
Guess the height of each object.
Then measure to check.

2.

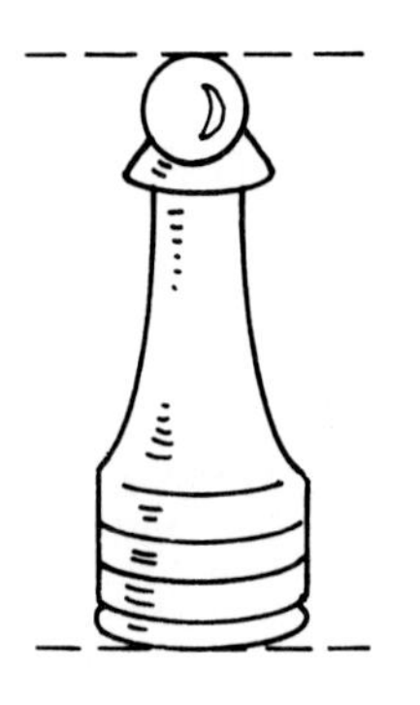

Guess: ________ cm

Measure: ______ cm

3.

Guess: ________ cm

Measure: ________cm

Name

1. Ring the objects that are longer than a meter.

2. Ring the objects that are taller than a meter.

Use a centimeter ruler.
About how tall is each object?
First, guess. Then measure to check.

3.

Guess: ________ cm

Measure: ______ cm

4.

Guess: ________ cm

Measure: ______ cm

Name

Which unit would you use to measure each object?
Ring the better unit.

1.

inches feet

2.

meters centimeters

Use a centimeter ruler.
Find the length to the closer centimeter.

3.

The caterpillar is closer to ____ centimeters.

4.

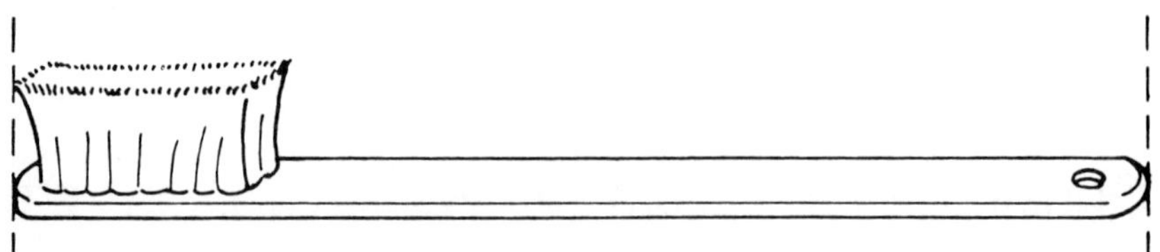

The toothbrush is closer to ____ centimeters.

Name

Use a centimeter ruler.
Find the length to the closer centimeter.

1.

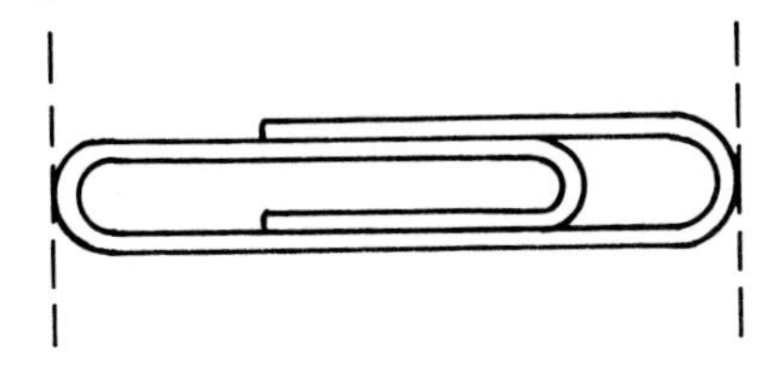

The paper clip is closer to ____ centimeters.

2.

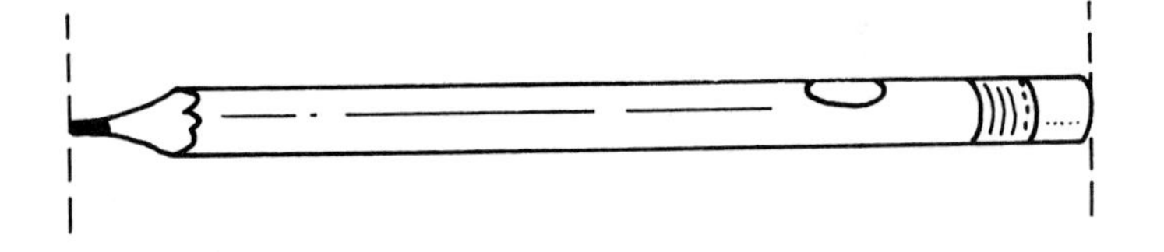

The pencil is closer to ____ centimeters.

3. Ring the objects that are longer than 1 meter.

4. Ring the objects that are taller than 1 meter.

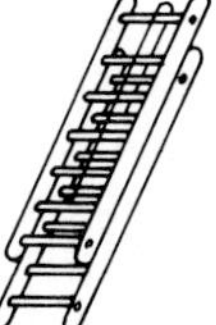

Name

Write how many tens and ones.
Then draw a line to show the trade.

1.

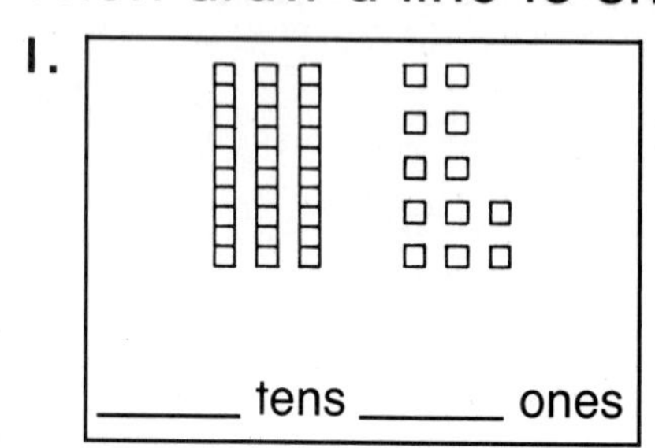

_____ tens _____ ones

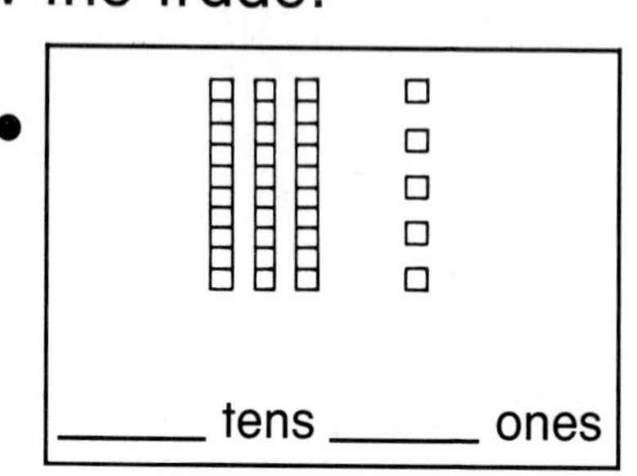

_____ tens _____ ones

2.

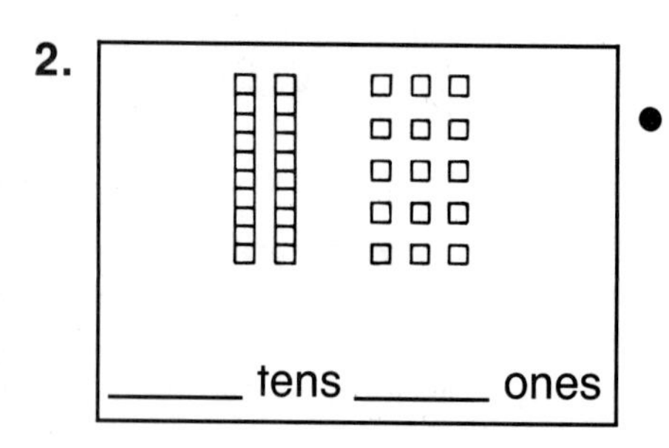

_____ tens _____ ones

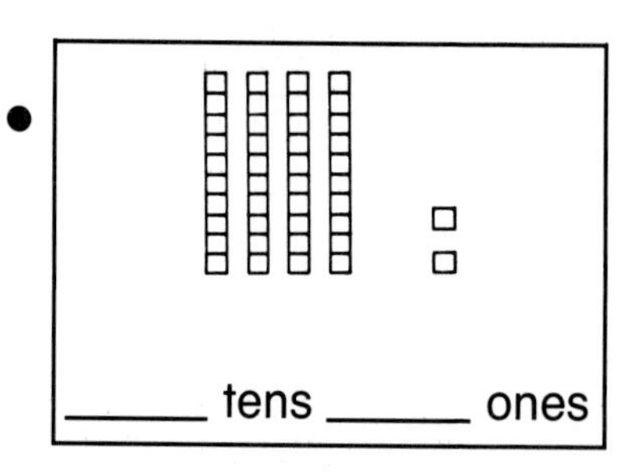

_____ tens _____ ones

Which unit would you use to measure each object? Ring the better unit.

3.

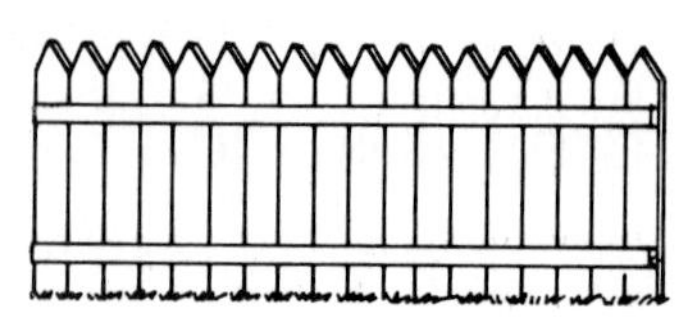

inches feet

4.

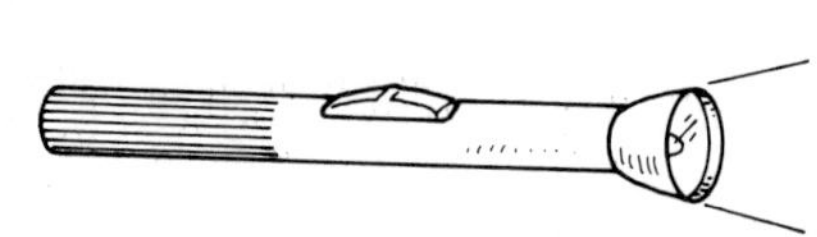

meters centimeters

Name ______________________________

Write how many in each group.
Ring another group of ten if you can.
Write how many in all.

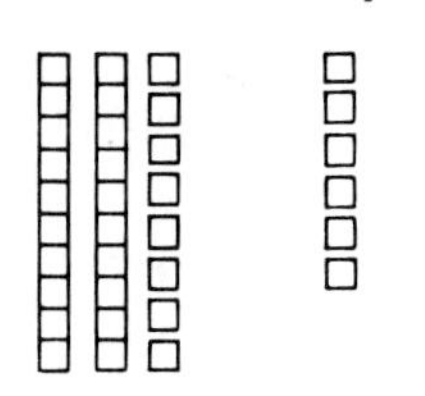

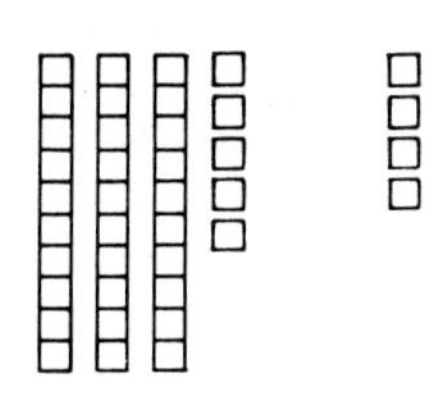

1. ______ and ______

______ in all

2. ______ and ______

______ in all

Are there enough ones to make a trade?
Ring a ten if you can.

Take this many.	Can you make a trade?	Write how many tens and ones.
3. □□□□ □□□□□□□□□□	______	______ ten ______ ones
4. □□□□□□□ □□□	______	______ ten ______ ones

Name

What is the number?
Complete the table.

1. Ben and Kate collect stamps.
Kate has 3 more stamps than Ben.

If Kate has	4	6	5	8	9
then Ben has					

Write how many in each group.
Then ring another group of ten if you can.
Write how many in all.

2. 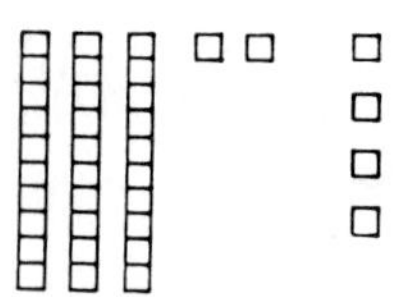

______ and ______

______ in all

3. 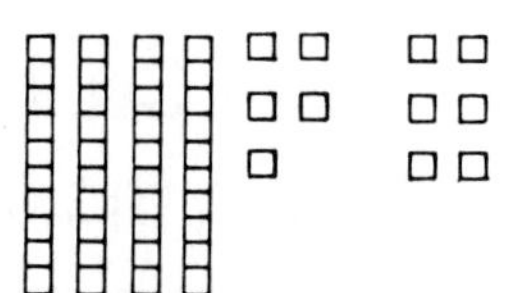

______ and ______

______ in all

Name

What is the number?
Complete the table.

1. Rusty and Toby are dogs.
Rusty is 2 years older than Toby.

When Toby is	2		5	
then Rusty will be		6		10

Write how many in each group.
Ring a ten if you can.
Write how many in all.

2.

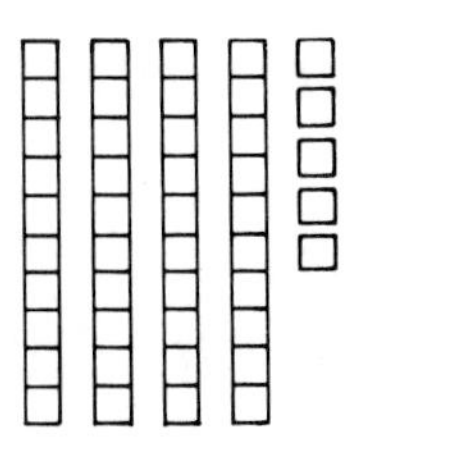

______ and ______

______ in all

3.

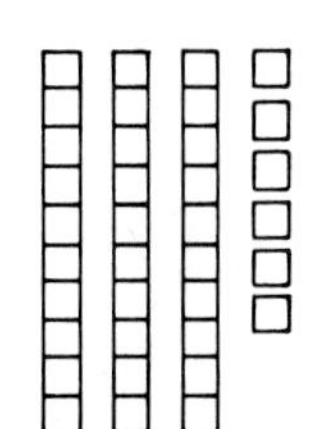

______ and ______

______ in all

Name

Write how many tens and ones.
Then draw a line to show each trade.

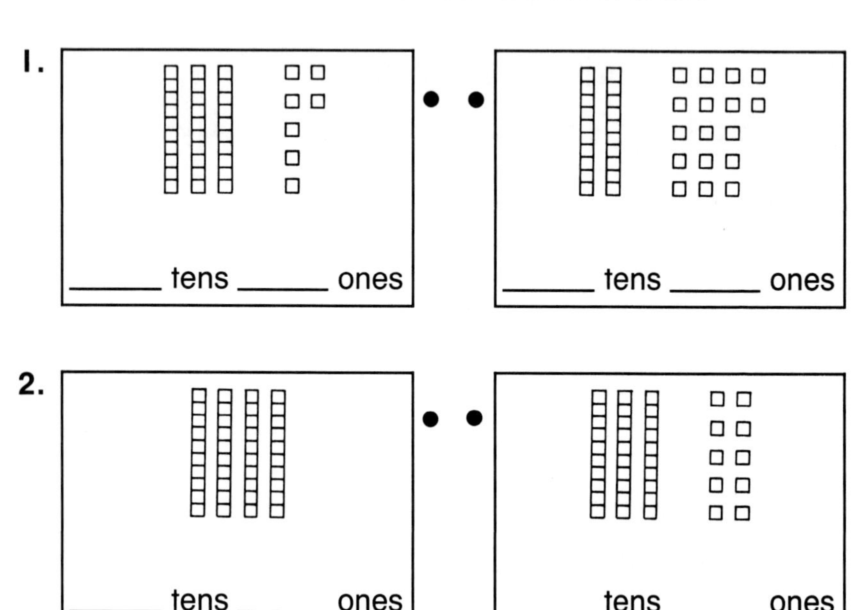

What is the number? Complete the table.

3. Allen and Patty paint pictures.
Allen paints 2 more pictures than Patty.

If Patty paints	2	5	1	3	4
then Allen paints					

Name

Decide if you need to trade.
Ring yes or no.

1. 35 take away 3

yes no

2. 52 take away 5

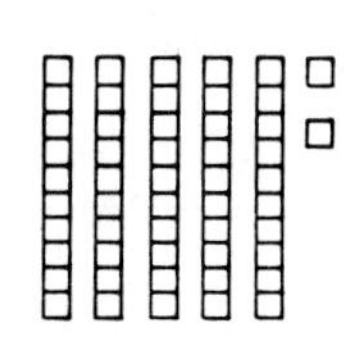

yes no

Write how many tens and ones.
Then draw a line to show each trade.

3.

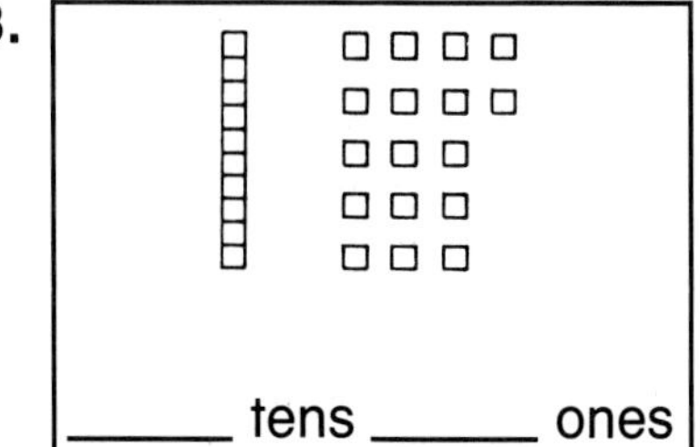

______ tens ______ ones

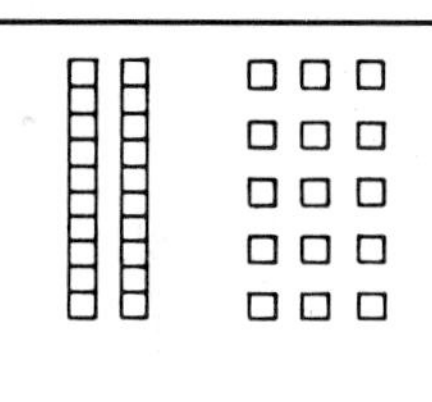

______ tens ______ ones

4.

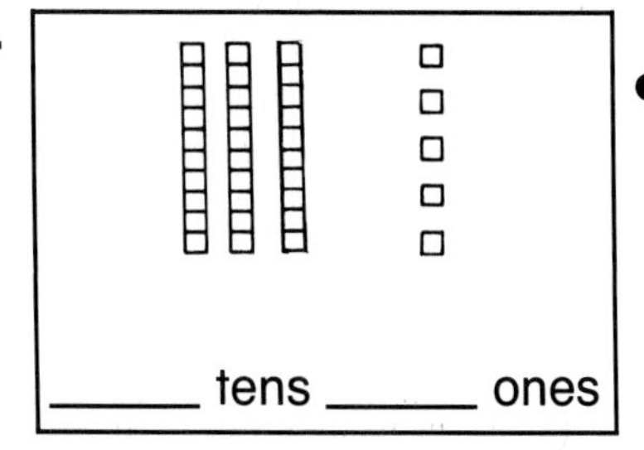

______ tens ______ ones

______ tens ______ ones

Name ______________________

Plant Sales = 5 plants	
Monday	
Wednesday	
Friday	

How many plants were sold each day?

1. Monday ______ 2. Friday ______

3. On which day were the least plants sold?

____

Decide if you need to trade.
Ring yes or no.

4. 24 take away 6

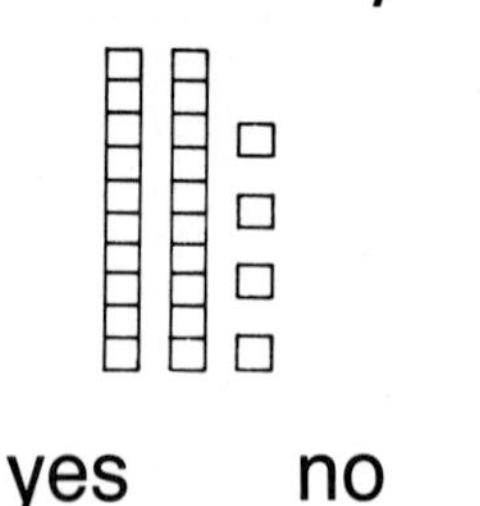

yes   no

5. 73 take away 2

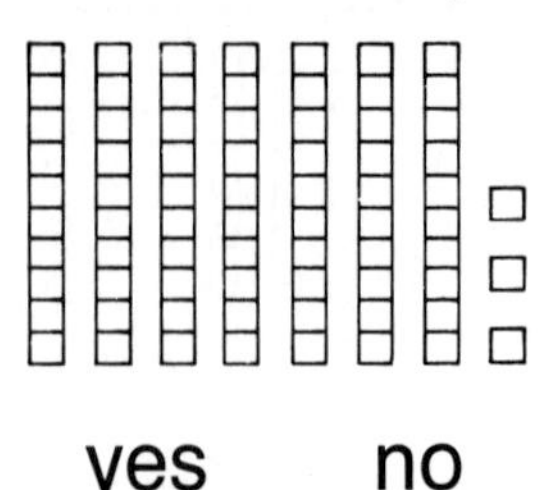

yes   no

Name

Decide if you need to trade.
Ring yes or no.

1. 32 take away 3

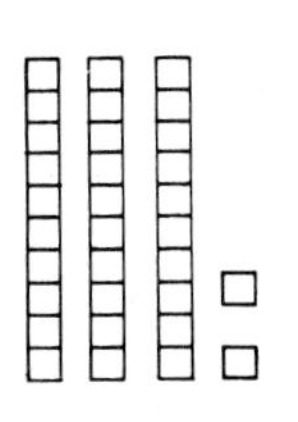

yes no

2. 57 take away 7

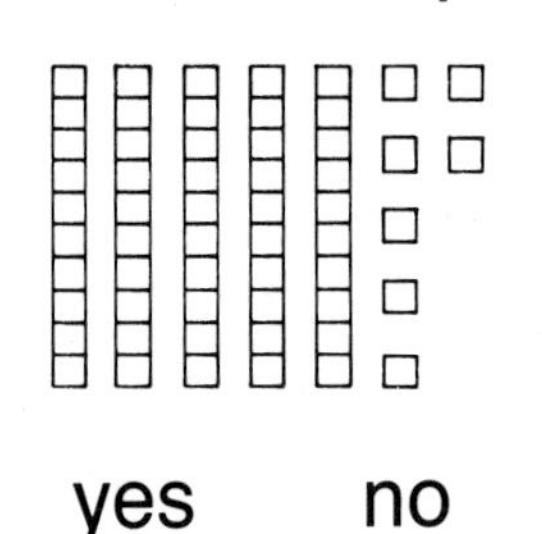

yes no

Book Sales = 10 books	
Monday	
Wednesday	
Friday	

How many books were sold each day?

3. Monday ______

4. Wednesday ______

5. On which day were the most books sold?

____

 Use before pages 203–204.

Name

Daily Review 66

Ring if you need to trade.
Then add.

1.

52	78	14	25
+6	+7	+9	+4

2.

49	58	18	43
+6	+4	+1	+5

Decide if you need to trade.
Ring yes or no.

3. 48 take away 2

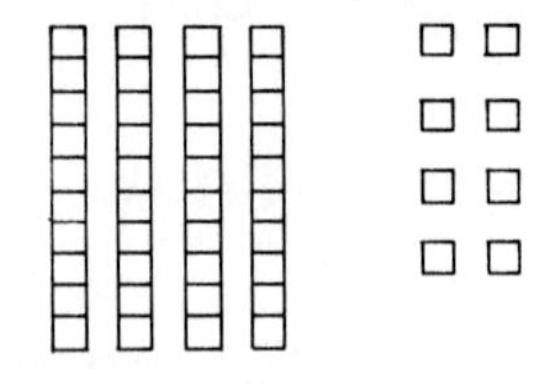

yes no

4. 26 take away 9

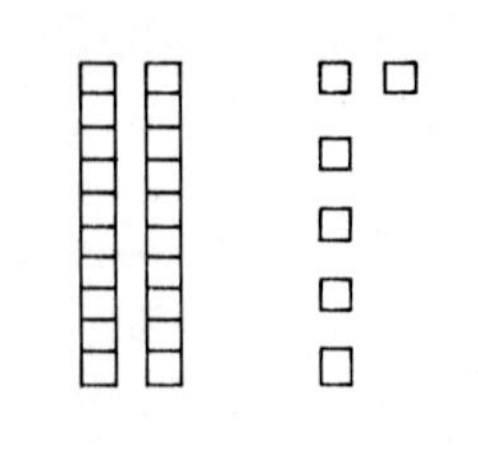

yes no

Name

Daily Review 67

Ring if you need to trade.
Then add.

1. $\begin{array}{r} 46 \\ +5 \\ \hline \end{array}$ $\begin{array}{r} 65 \\ +2 \\ \hline \end{array}$ $\begin{array}{r} 49 \\ +3 \\ \hline \end{array}$ $\begin{array}{r} 65 \\ +7 \\ \hline \end{array}$

2. $\begin{array}{r} 22 \\ +8 \\ \hline \end{array}$ $\begin{array}{r} 47 \\ +2 \\ \hline \end{array}$ $\begin{array}{r} 26 \\ +5 \\ \hline \end{array}$ $\begin{array}{r} 53 \\ +4 \\ \hline \end{array}$

Minutes Run	 = 2 minutes
Jan	
Bob	
Tim	

3. Who ran the longest time? ________

4. How many minutes did Jan run? ______

Name

Daily Review 68

Ring if you need to trade. Add.

1. $\begin{array}{r} 36 \\ +24 \\ \hline \end{array}$ $\begin{array}{r} 48 \\ +13 \\ \hline \end{array}$ $\begin{array}{r} 73 \\ +22 \\ \hline \end{array}$ $\begin{array}{r} 19 \\ +12 \\ \hline \end{array}$

2. $\begin{array}{r} 43 \\ +18 \\ \hline \end{array}$ $\begin{array}{r} 67 \\ +21 \\ \hline \end{array}$ $\begin{array}{r} 35 \\ +25 \\ \hline \end{array}$ $\begin{array}{r} 35 \\ +38 \\ \hline \end{array}$

Ring if you need to trade. Then add.

3. $\begin{array}{r} 69 \\ +\ 4 \\ \hline \end{array}$ $\begin{array}{r} 28 \\ +\ 7 \\ \hline \end{array}$ $\begin{array}{r} 12 \\ +\ 4 \\ \hline \end{array}$ $\begin{array}{r} 65 \\ +\ 4 \\ \hline \end{array}$

4. $\begin{array}{r} 45 \\ +\ 6 \\ \hline \end{array}$ $\begin{array}{r} 87 \\ +\ 3 \\ \hline \end{array}$ $\begin{array}{r} 36 \\ +\ 3 \\ \hline \end{array}$ $\begin{array}{r} 92 \\ +\ 7 \\ \hline \end{array}$

Name

Ring if you need to trade.
Then find the sum.

1.

50	47	38	29
+ 8	+ 22	+ 59	+ 11

2.

64	48	13	25
+ 29	+ 35	+ 55	+ 24

Ring if you need to trade. Then add.

3.

25	64	18	39
+ 4	+ 8	+ 7	+ 3

4.

77	42	54	82
+ 5	+ 6	+ 4	+ 9

Name ______________________________

Write the sum. Write the ¢.

1.

25¢	46¢	38¢	63¢
+ 32¢	+ 27¢	+ 17¢	+ 11¢

2.

11¢	52¢	70¢	38¢
+ 86¢	+ 19¢	+ 21¢	+ 58¢

Ring if you need to trade.
Then add.

3.

26	72	28	48
+ 14	+ 12	+ 31	+ 25

4.

53	64	79	28
+ 21	+ 27	+ 18	+ 47

Name

60¢

46¢

28¢

37¢

Look at the picture. Solve.

1. If you buy a goldfish and a net, what is the cost altogether?

____ ¢
☐ ____ ¢
______
¢

2. Joe buys a bowl and fish food. How much does he spend?

____ ¢
☐ ____ ¢
______
¢

Ring if you need to trade. Then add.

3.

$$\begin{array}{r} 74 \\ +12 \\ \hline \end{array} \qquad \begin{array}{r} 63 \\ +31 \\ \hline \end{array} \qquad \begin{array}{r} 62 \\ +29 \\ \hline \end{array} \qquad \begin{array}{r} 35 \\ +31 \\ \hline \end{array}$$

Name

Estimate the sum.
Ring more or less.

1.

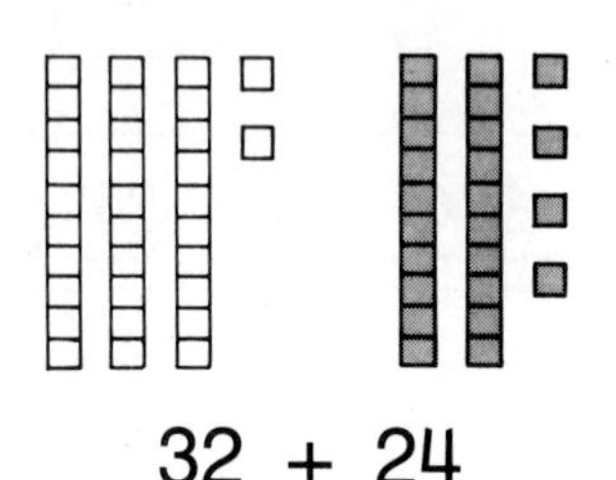

32 + 24

more than 60

less than 60

2.

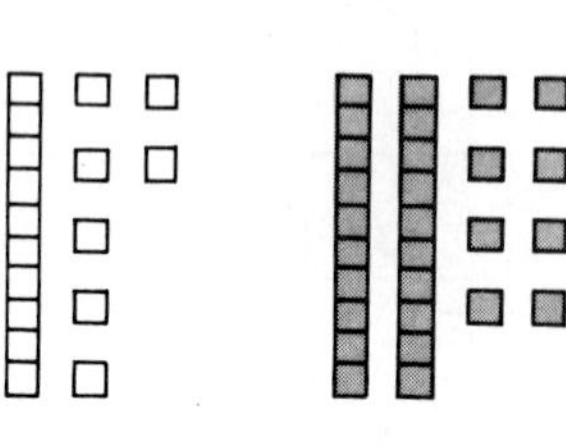

17 + 28

more than 40

less than 40

Write the sums. Write the ¢.

3.

42¢	62¢	67¢	34¢
+ 30¢	+ 13¢	+ 29¢	+ 10¢

Ring if you need to trade. Add.

4.

28	62	46	37
+ 19	+ 21	+ 31	+ 29

Name ______________________

Write the sum.
Look for a pattern.

1.

35	45	55	65
+ 9	+ 9	+ 9	+ 9

2.

34	44	54	64
+ 4	+ 4	+ 4	+ 4

Ring if you need to trade. Then add.

3.

41	62	57	32
+36	+19	+16	+35

4.

54	73	34	41
+35	+19	+23	+29

  Use before pages 221–222.

Name

Daily Review 74

Write the sums.

1.

42	24	31	46
31	26	19	20
+ 18	+ 12	+ 26	+ 24

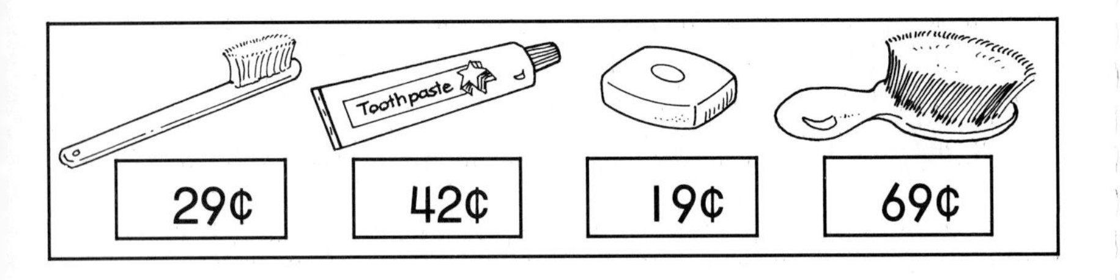

Look at the picture. Solve.

2. How much do a toothbrush and a tube of toothpaste cost?

____ ¢
☐ ____ ¢
¢

3. Peg buys a bar of soap and a hairbrush. How much does she spend?

____ ¢
☐ ____ ¢
¢

Name ______________________

Is there enough information?
Ring enough or not enough.

1. Jerry collected 7 eggs from his two chickens. The red hen laid 4 eggs. How many eggs did the black hen lay?

enough not enough

2. The horse ate 3 apples. The pig ate more apples than the horse. How many apples did the pig eat?

enough not enough

Write the sum.

3.

33	19	24	18
21	24	9	23
+ 27	+ 15	+ 31	+ 7

Name ____________________

Is there enough information?
Ring enough or not enough.

1. Sarah had 22 marbles. She gave 9 marbles to her brother, Paul. How many marbles does Sarah have left?

enough   not enough

2. Paul has 13 postcards. Sarah has fewer postcards than Paul. How many postcards does Sarah have?

enough   not enough

Write the sums.

3.

21	12	42	15
42	12	31	18
+ 14	+ 12	+ 12	+ 24

Name

Do you need to trade?
Ring yes or no. Then subtract.

1.

74 Yes	38 Yes	52 Yes
− 6 No	− 6 No	− 8 No

2.

25 Yes	96 Yes	46 Yes
− 7 No	− 4 No	− 7 No

Cross out the facts you do not need.

3. Bill read 17 books about horses, 13 books about dogs, and 9 books about people. How many more books about horses than people did he read?

Write the sums.

4.

14	12	68	51
23	15	11	25
+22	+19	+10	+14

Name

Do you need to trade?
Ring yes or no.
Then write the difference.

1. $\begin{array}{r} 45 \\ -\ 8 \\ \hline \end{array}$ Yes No $\begin{array}{r} 28 \\ -\ 3 \\ \hline \end{array}$ Yes No $\begin{array}{r} 67 \\ -\ 9 \\ \hline \end{array}$ Yes No

2. $\begin{array}{r} 58 \\ -\ 9 \\ \hline \end{array}$ Yes No $\begin{array}{r} 43 \\ -\ 7 \\ \hline \end{array}$ Yes No $\begin{array}{r} 66 \\ -\ 4 \\ \hline \end{array}$ Yes No

Is there enough information?
Ring enough or not enough.

3. Reggie ate 8 grapes. Len ate more grapes than Reggie. How many grapes did Len eat?

enough not enough

Name

Ring if you need to trade.
Write the difference.

1.

81	75	39	57
− 63	− 49	− 27	− 49

2.

42	58	63	44
− 34	− 14	− 11	− 27

Do you need to trade?
Ring yes or no.
Then write the difference.

3.

63	Yes	42	Yes	48	Yes
− 9	No	− 5	No	− 3	No

  Use before pages 243–244.

Name

Ring if you need to trade.
Write the difference.

1. $\begin{array}{r} 40 \\ -18 \\ \hline \end{array}$ $\begin{array}{r} 69 \\ -30 \\ \hline \end{array}$ $\begin{array}{r} 70 \\ -26 \\ \hline \end{array}$ $\begin{array}{r} 80 \\ -20 \\ \hline \end{array}$

2. $\begin{array}{r} 39 \\ -25 \\ \hline \end{array}$ $\begin{array}{r} 94 \\ -37 \\ \hline \end{array}$ $\begin{array}{r} 42 \\ -36 \\ \hline \end{array}$ $\begin{array}{r} 63 \\ -41 \\ \hline \end{array}$

Do you need to trade?
Ring yes or no. Then write the difference.

3. $\begin{array}{r} 37 \\ -\ 2 \\ \hline \end{array}$ Yes No $\begin{array}{r} 35 \\ -\ 9 \\ \hline \end{array}$ Yes No $\begin{array}{r} 26 \\ -\ 8 \\ \hline \end{array}$ Yes No

4. $\begin{array}{r} 25 \\ -\ 8 \\ \hline \end{array}$ Yes No $\begin{array}{r} 44 \\ -\ 3 \\ \hline \end{array}$ Yes No $\begin{array}{r} 63 \\ -\ 7 \\ \hline \end{array}$ Yes No

Name

Decide if you need to add or subtract. Then solve.

1. At lunch, 86 pieces of fruit were served. There were 49 apples. the rest were pears. How many pears were served?

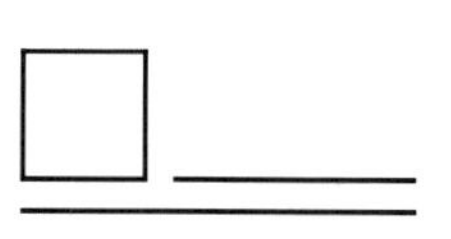

______ pears

Ring if you need to trade. Write the difference.

2. $\begin{array}{r} 74 \\ -59 \\ \hline \end{array}$ $\begin{array}{r} 87 \\ -38 \\ \hline \end{array}$ $\begin{array}{r} 25 \\ -19 \\ \hline \end{array}$ $\begin{array}{r} 65 \\ -34 \\ \hline \end{array}$

3. $\begin{array}{r} 29 \\ -13 \\ \hline \end{array}$ $\begin{array}{r} 35 \\ -28 \\ \hline \end{array}$ $\begin{array}{r} 57 \\ -32 \\ \hline \end{array}$ $\begin{array}{r} 60 \\ -34 \\ \hline \end{array}$

Name

Daily Review 82

Subtract.

1.

48¢	60¢	92¢	34¢
− 29¢	− 36¢	− 6¢	− 31¢

Decide if you need to add or subtract. Then solve.

2. Monica and Saleem saw 19 cows on their trip to the farm. They saw 23 more chickens than cows. How many chickens did they see?

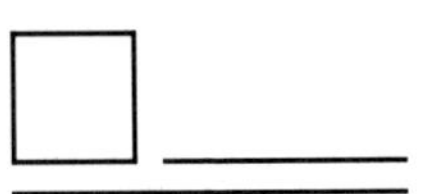

______ chickens

Ring if you need to trade.
Write the difference.

3.

60	39	80	30
− 17	− 20	− 46	− 12

Name

Estimate the difference.
Ring more than or less than.

1.

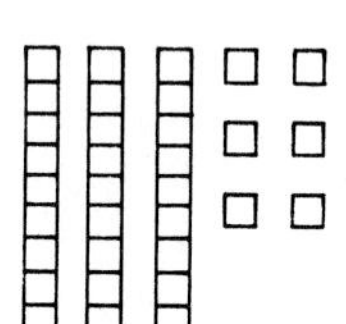

36 − 22

more than 10

less than 10

2.

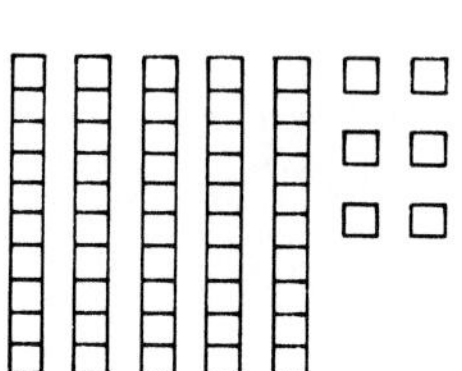

56 − 39

more than 20

less than 20

Subtract.

3.

63¢	87¢	25¢	48¢
− 38¢	− 46¢	− 19¢	− 29¢

Ring if you need to trade.
Write the difference.

4.

90	80	50	49
− 24	− 60	− 36	− 10

Use before pages 253–254.

Name

Write the difference.
Look for a pattern.

1. $\begin{array}{r} 34 \\ -9 \\ \hline \end{array}$ $\begin{array}{r} 44 \\ -9 \\ \hline \end{array}$ $\begin{array}{r} 54 \\ -9 \\ \hline \end{array}$ $\begin{array}{r} 64 \\ -9 \\ \hline \end{array}$

2. $\begin{array}{r} 73 \\ -4 \\ \hline \end{array}$ $\begin{array}{r} 63 \\ -4 \\ \hline \end{array}$ $\begin{array}{r} 53 \\ -4 \\ \hline \end{array}$ $\begin{array}{r} 43 \\ -4 \\ \hline \end{array}$

Decide if you need to add or subtract. Then solve.

3. 24 birds sing in the trees. 42 birds look for worms on the ground. How many birds are in the park?

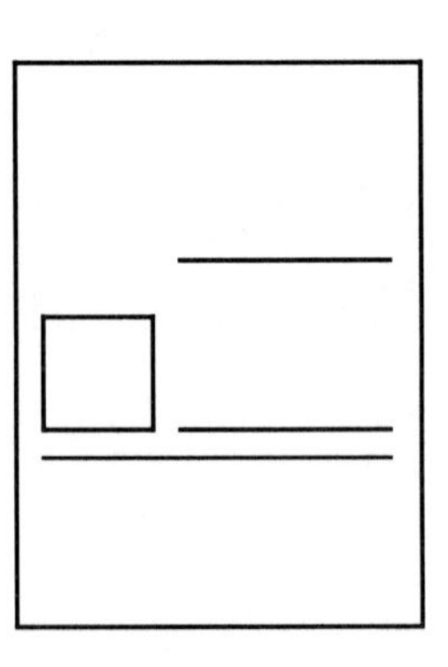

______ birds

Name

Daily Review 85

Subtract.
Then add to check.

1.
$$\begin{array}{r} 43 \\ -\,28 \\ \hline \end{array} \qquad \begin{array}{r} \_\_\_ \\ +\ \_\_\_ \\ \hline \end{array}$$

2.
$$\begin{array}{r} 75 \\ -\,39 \\ \hline \end{array} \qquad \begin{array}{r} \_\_\_ \\ +\ \_\_\_ \\ \hline \end{array}$$

3.
$$\begin{array}{r} 59 \\ -\,36 \\ \hline \end{array} \qquad \begin{array}{r} \_\_\_ \\ +\ \_\_\_ \\ \hline \end{array}$$

4.
$$\begin{array}{r} 64 \\ -\,27 \\ \hline \end{array} \qquad \begin{array}{r} \_\_\_ \\ +\ \_\_\_ \\ \hline \end{array}$$

Estimate the difference.
Ring more than or less than.

5.

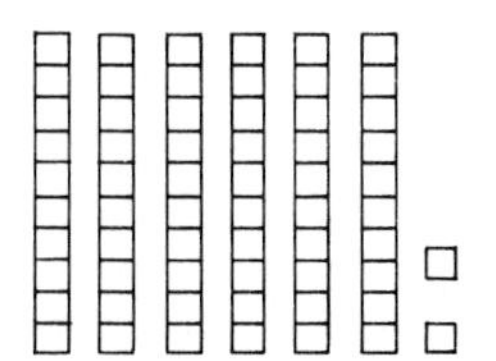

62 – 38

more than 30

less than 30

6.

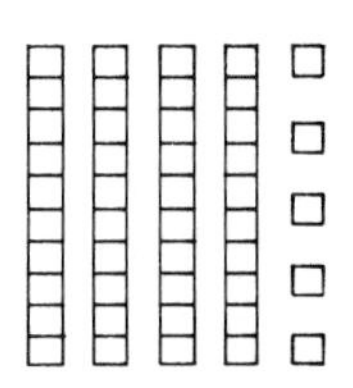

45 – 21

more than 20

less than 20

Name

Show each step. Then solve.

1. Karen had 59¢. She earned 35¢ raking leaves. How much does she have now?

Karen spent 18¢ to buy a magic trick. How much does she have left?

Step 1	Step 2
____¢ ☐ ____¢ ______	____¢ ☐ ____¢ ______

Estimate the difference.
Ring more than or less than.

2. 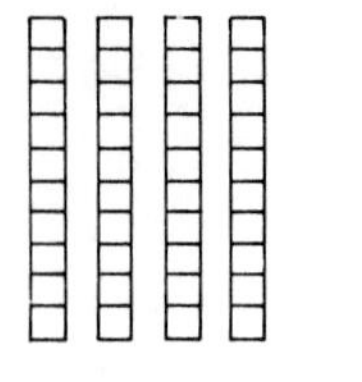

48 − 15

more than 30

less than 30

3. 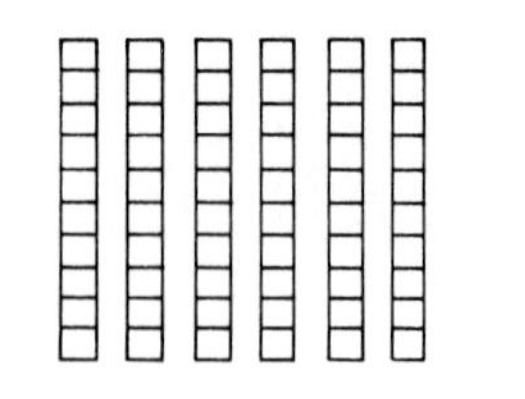

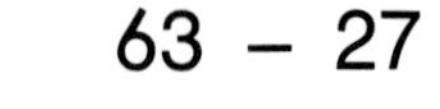

63 − 27

more than 40

less than 40

Name ______

Write how many.

1.

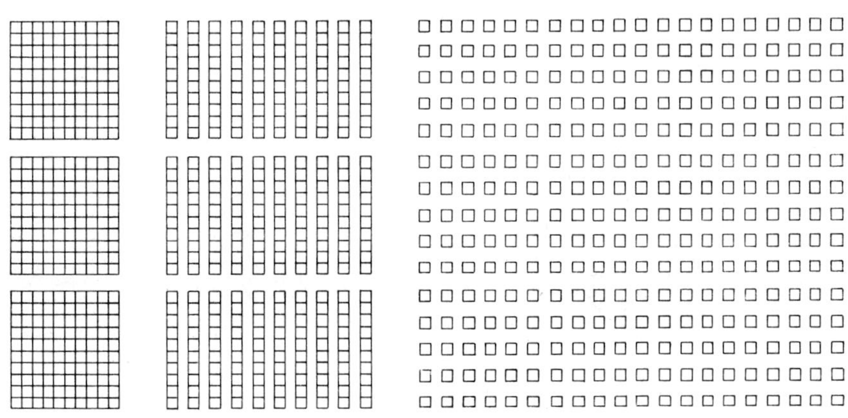

____ hundreds = ____ tens = ____ ones

Subtract.
Add to check.

2.

$$\begin{array}{r} 68 \\ -\,37 \\ \hline \end{array} \qquad \begin{array}{r} \_\_\_ \\ +\,\_\_\_ \\ \hline \end{array}$$

3.

$$\begin{array}{r} 41 \\ -\,19 \\ \hline \end{array} \qquad \begin{array}{r} \_\_\_ \\ +\,\_\_\_ \\ \hline \end{array}$$

Name

Daily Review 88

Subtract.

1.

$$\begin{array}{r} 44 \\ -9 \\ \hline \end{array} \quad \begin{array}{r} 45 \\ -9 \\ \hline \end{array} \quad \begin{array}{r} 46 \\ -9 \\ \hline \end{array} \quad \begin{array}{r} 47 \\ -9 \\ \hline \end{array}$$

2.

$$\begin{array}{r} 71 \\ -9 \\ \hline \end{array} \quad \begin{array}{r} 72 \\ -9 \\ \hline \end{array} \quad \begin{array}{r} 73 \\ -9 \\ \hline \end{array} \quad \begin{array}{r} 74 \\ -9 \\ \hline \end{array}$$

Show each step. Then solve.

3. A tape costs 79¢. Toby has 37¢. How much more does Toby need? Toby's sister gives her 22¢. How much does she still need?

Step 1
____¢   □ ____¢   ______

Step 2
____¢   □ ____¢   ______

Name ____________________

Write the number.

1. 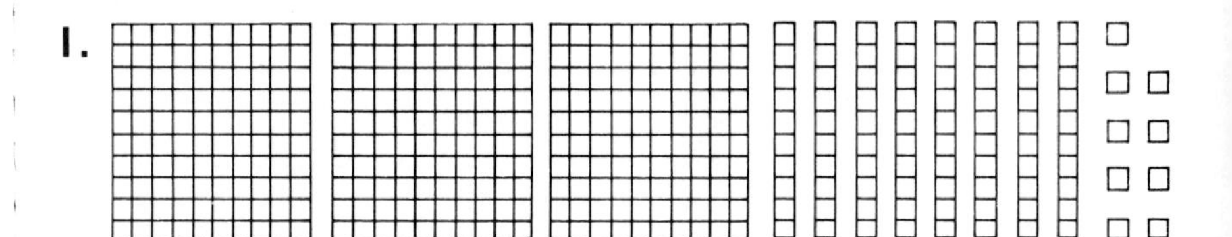

____

2.

____

Write how many.

3. 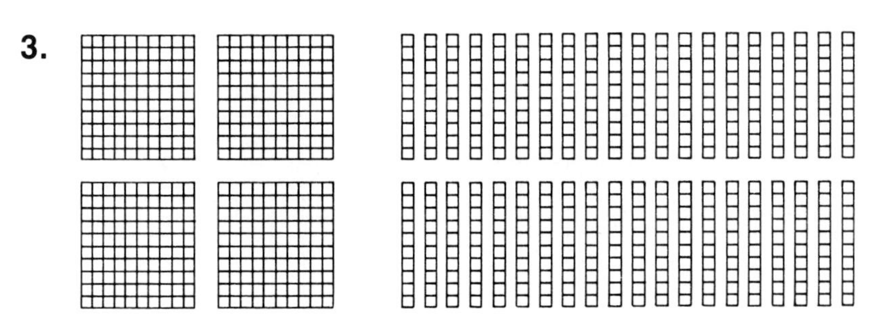

____ hundreds = _____ tens

Name ______________________

How many?

1. 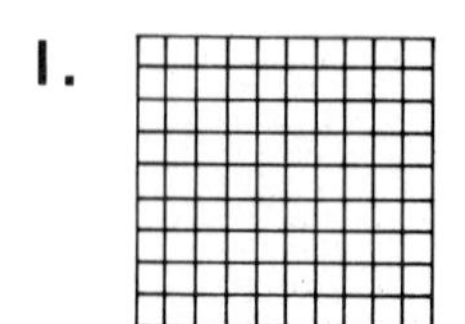 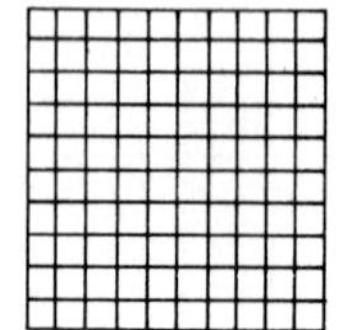 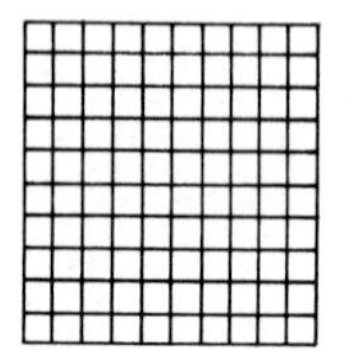 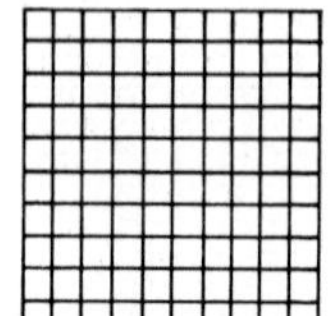

_____ hundreds _____ tens _____ ones

________ and _____ and _____

____

Write the number.

2. 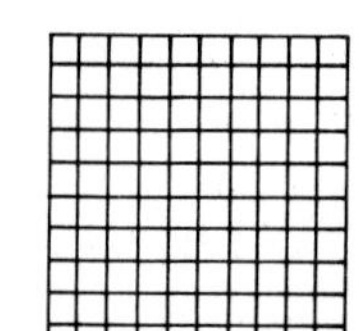  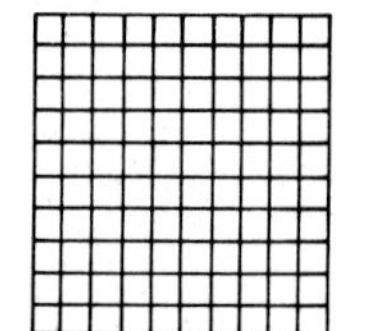

____

3. 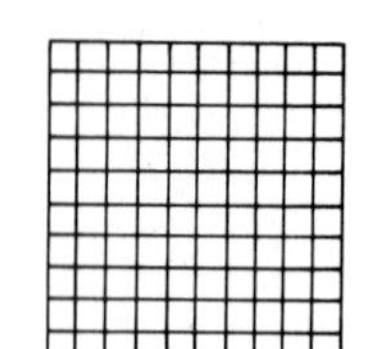 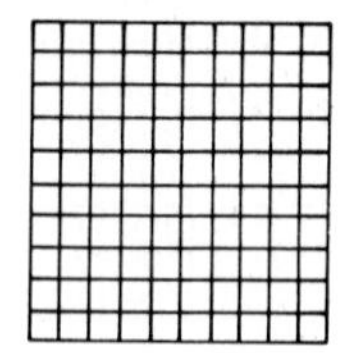 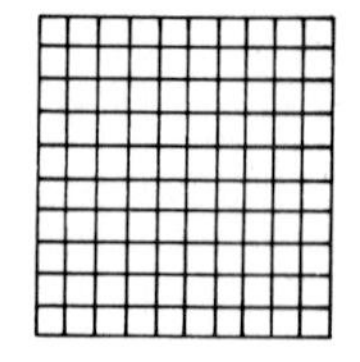 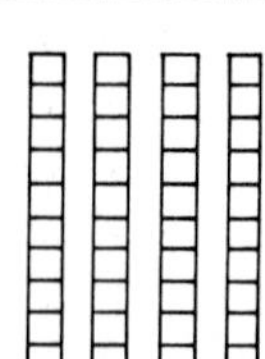

____

Name

Daily Review 91

Read the problem.
Underline the question. Ring the facts.

1. Fran has 47¢. She earns 19¢ more.
How much does she have now?

Ring + or − .
Then solve.

2. Fran has 47¢.
She earns 19¢ more.
How much does she have now?

+
−

Write the answer in the sentence.
Does the answer make sense? Ring yes or no.

3. Fran has ________ now.

yes no

Write the number.

4.
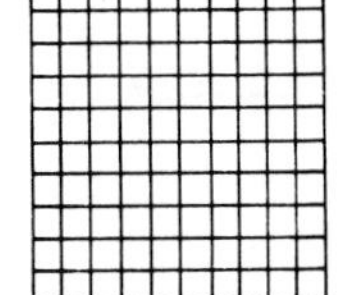
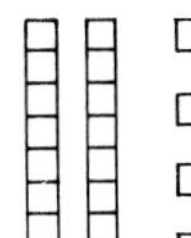

____

Use before pages 279–280.

Name

Daily Review 92

Write the missing numbers.

1.

After	
388	
457	
699	
864	

2.

Before	
	301
	530
	641
	440

3.

Between		
438		440
310		312
898		900
599		601

How many?

4. 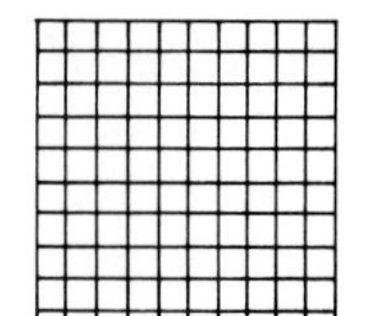 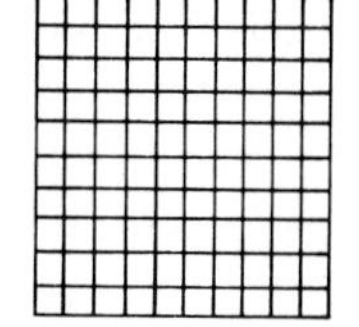 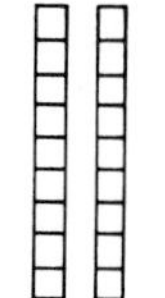

_____ hundreds _____ tens _____ ones

________ and _____ and _____

____

Name

Write > or < in each circle.

1. 479 ◯ 420   2. 245 ◯ 230

Read the problem.
Underline the question.
Ring the facts.

3. Roger needs 62¢. He has 48¢.
How much more does he need?

Ring + or −.
Then solve.

4. Roger needs 62¢.
He has 48¢. How much
more does he need?

+
−

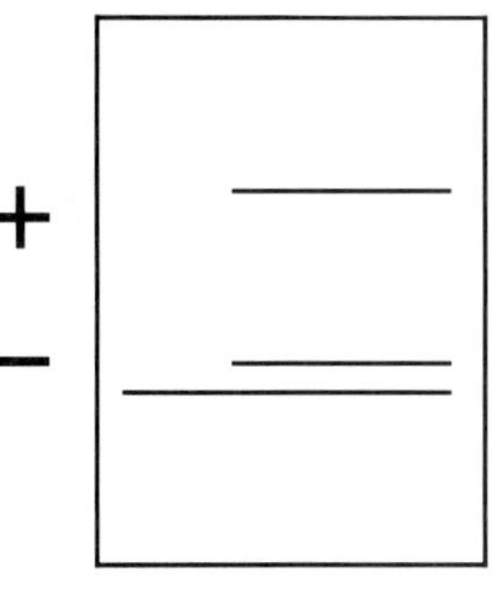

Write the answer in the sentence.
Does the answer make sense? Ring yes or no.

4. Roger needs ______¢.

yes   no

Name ____________________

Write 100 less and 100 more for each number.

1. _____ 239 _____

2. _____ 697 _____

Write the missing numbers.

3.

After	
765	
679	
321	
310	
799	
548	

4.

Before	
	460
	220
	131
	625
	397
	470

5.

Between		
234		236
449		451
678		680
583		585
851		853
199		201

Use before pages 285–286.

Name ______________________

Daily Review 95

Write the numbers.

1. three hundred forty-six ______
2. six hundred eighteen ______
3. one hundred ninety-nine ______

Write 100 less and 100 more for each number.

4. ____ 476 ____
5. ____ 891 ____
6. ____ 233 ____
7. ____ 654 ____

Write $>$ or $<$ in the circle.

8. 398 ◯ 401
9. 563 ◯ 536
10. 635 ◯ 827
11. 953 ◯ 924

Name

1. What choices do you have?
Color the chart.

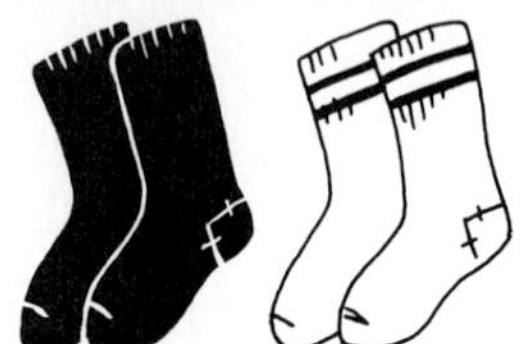

Green Yellow

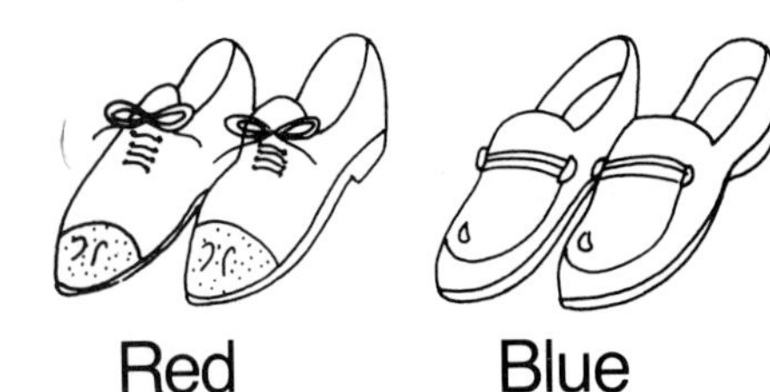

Red Blue

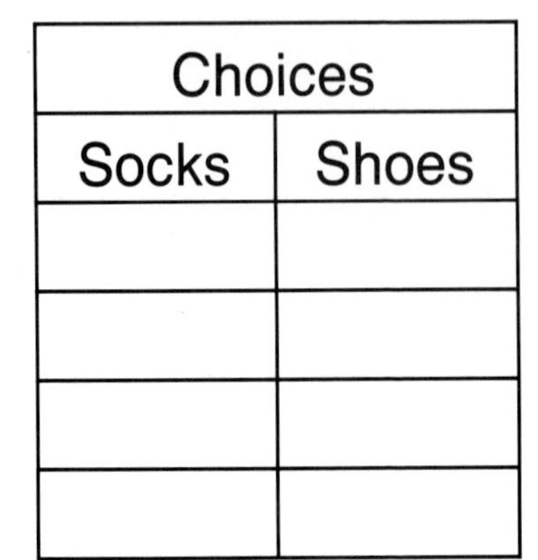

Choices	
Socks	Shoes

Write the numbers.

2. four hundred thirteen ________

3. two hundred forty-six ________

4. seven hundred two ________

Write > or <.

5. 263 ◯ 189

6. 463 ◯ 364

Name ______________________

Daily Review 97

Write the amount.

1.

______

2.

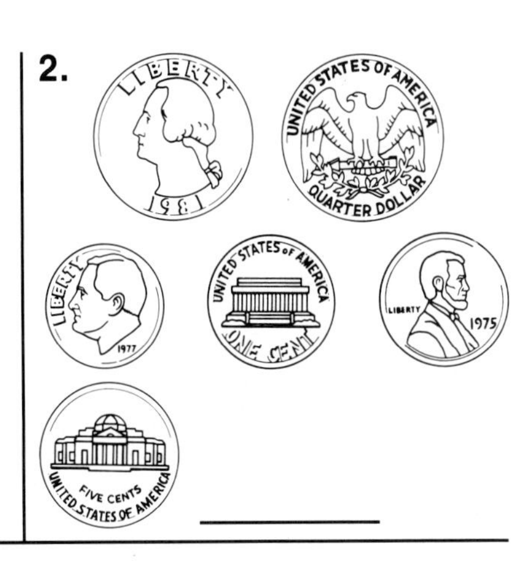

______

3. What choices do you have?
Color the chart.

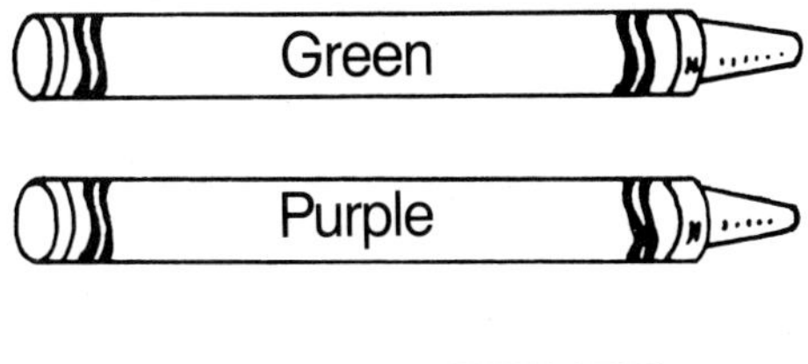

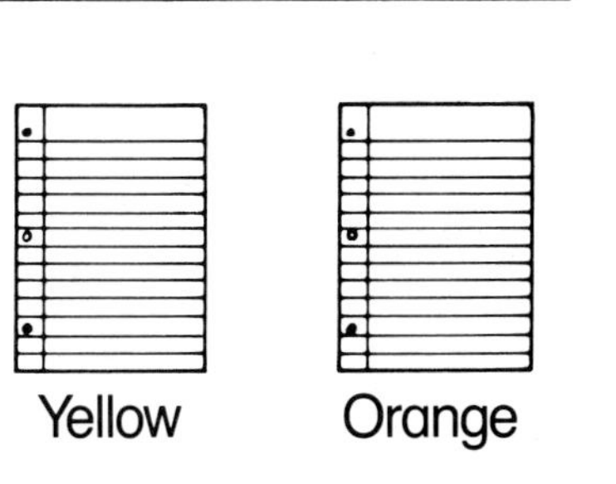

Choices	
Crayon	Paper

Name ____________________

Does the amount equal exactly a half dollar?
Write yes or no.

1.

____

2.

____

Write the amount.

3.

____

4.

____

Write the numbers.

5. two hundred two __________

6. nine hundred ninety-nine __________

Name

Write the amount.

1.

____

2.

____

Count on to find how much.

3.

____

4.

____

 Use before pages 303–304.

Name

Write yes or no.

What You Buy.	You Give	Will you get change?

1.

____

2.

____

Does the amount equal exactly a half dollar?
Write yes or no.

3.

____

Name

1. Ring the two items that cost 58¢.

2. Ring the two items that cost 73¢.

Write the amount.

3.

____

  Use before pages 309–310.

Name

Does the bank contain exactly $1.00?
Write yes or no.

1.

______

Write yes or no.

What You Buy.	You Give	Will you get change?

2.

______

Name

Ring the money to show the amount.

3. Ring the two items that cost 61¢.

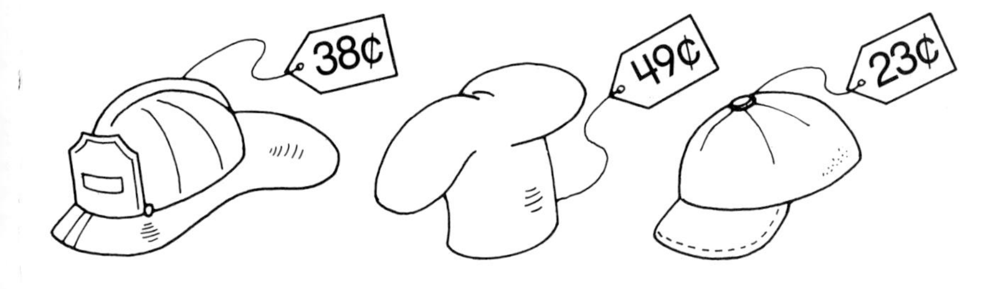

  Use before pages 313–314.

Name

Write the amount.

1.

____

2.

____

Does the bank contain exactly $1.00?
Write yes or no.

3.

____

Name

Count on to find how much.

Ring the money to show the amount.

Name

Write how much money.
Then ring the lesser amount.

1.

______

2.

______

Write the amount.

3.

______

Name

Ring the most sensible answer.
Then tell why your choice
makes sense.

1. About how many times do second graders brush their teeth in a day?

3 10 30

Why does your answer make sense?

____

____

Count on to find how much.

2.

______ ______ ______ ______ ______ in all

Name

Color and complete.

1\.

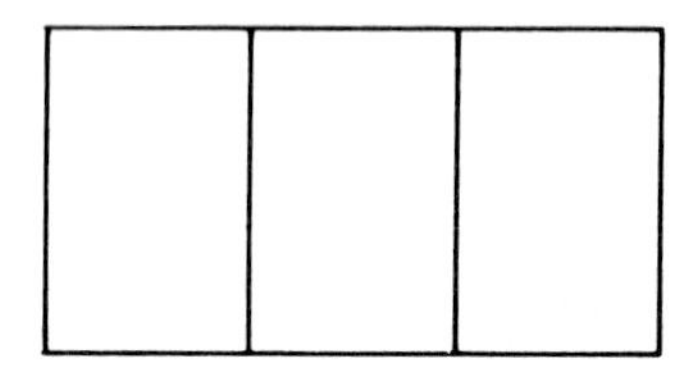

$\frac{2}{3}$ 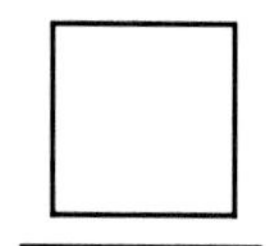

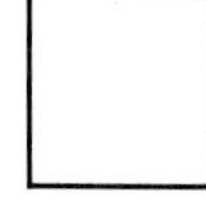

☐ red parts

☐ equal parts

Write how much money.
Then ring the lesser amount.

3\.

4\.

Use before pages 331–332.

Name ______________________

Complete.

1. Color $\frac{3}{5}$ green.

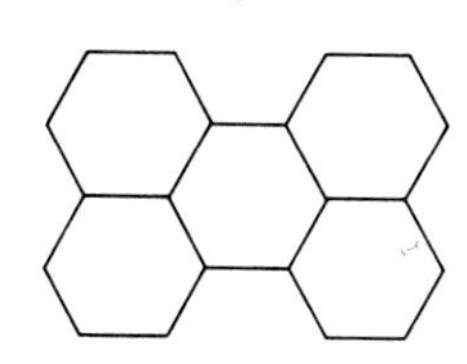

green parts

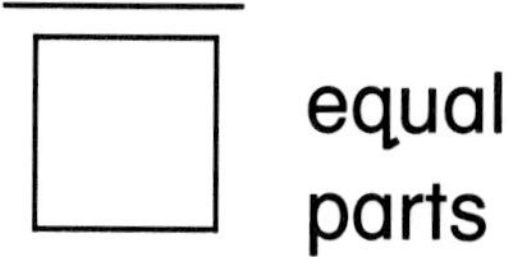

equal parts

2. Color $\frac{4}{6}$ red.

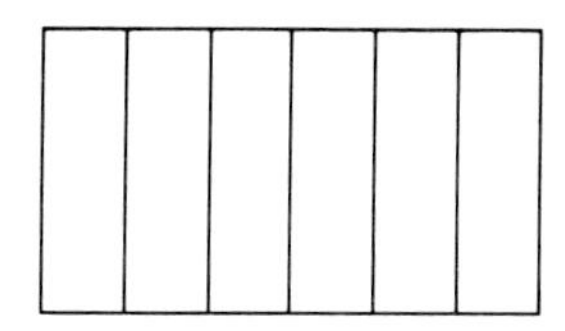

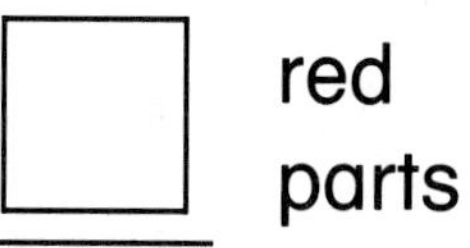

red parts

equal parts

Ring the most sensible answer.
Then tell why your choice makes sense.

3. About how many bowls of cereal does a second grader eat in a week?

6   30   120

Why does your answer make sense?

____

____

 Use before pages 333–334.

Name

Daily Review 110

Complete.

1. Color $\frac{2}{5}$ yellow.

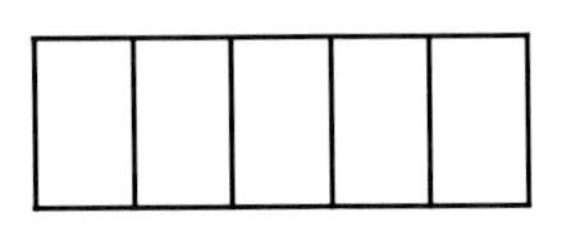

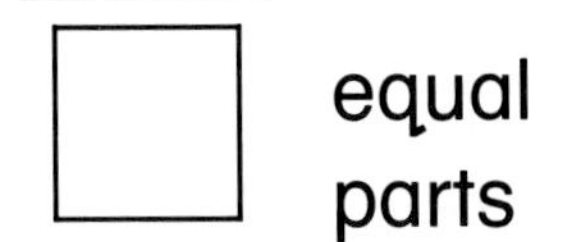

yellow parts

equal parts

2. Color $\frac{5}{6}$ green.

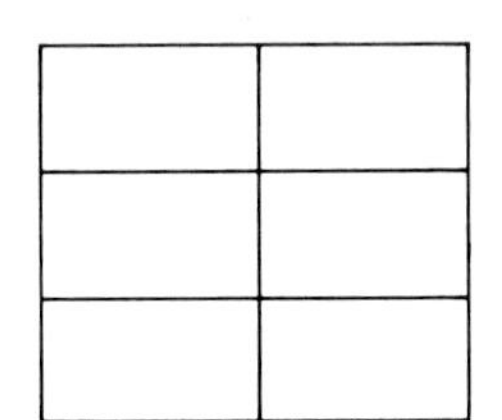

green parts

equal parts

3. Color $\frac{3}{3}$ yellow.

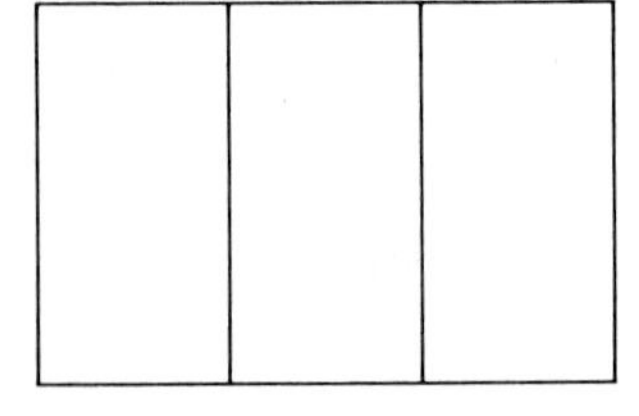

yellow parts

equal parts

4. Color $\frac{1}{2}$ green.

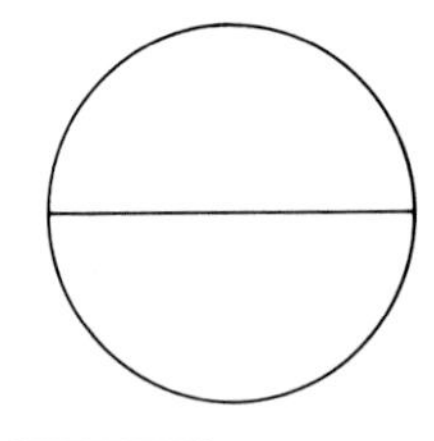

green parts

equal parts

Name ______________________

Write the fraction.

1. What fraction is black?

black cats

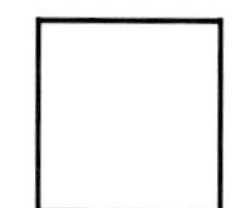

cats in all

Complete.

2. Color $\frac{4}{6}$ blue.

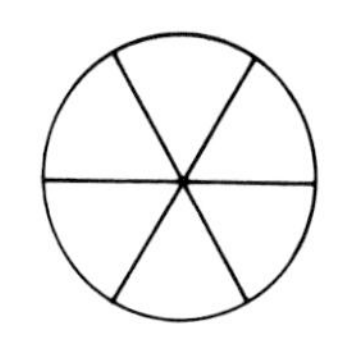

$\frac{4}{6}$ 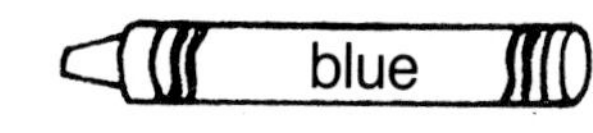

blue parts

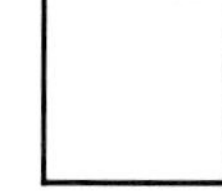

equal parts

3. Color $\frac{1}{4}$ green.

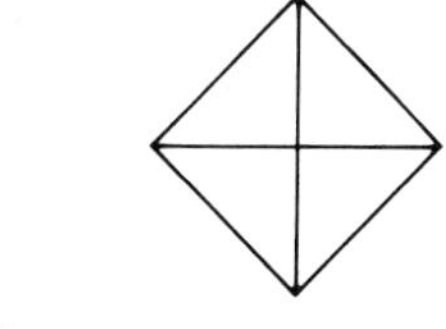

$\frac{1}{4}$ green

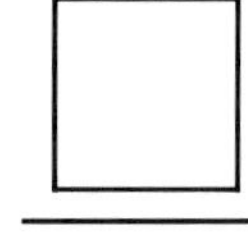

green parts

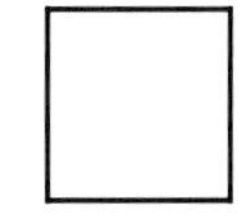

equal parts

Name

Draw a picture. Then solve.

1. There are 3 red flowers and 1 yellow flower. What part of the group is red?

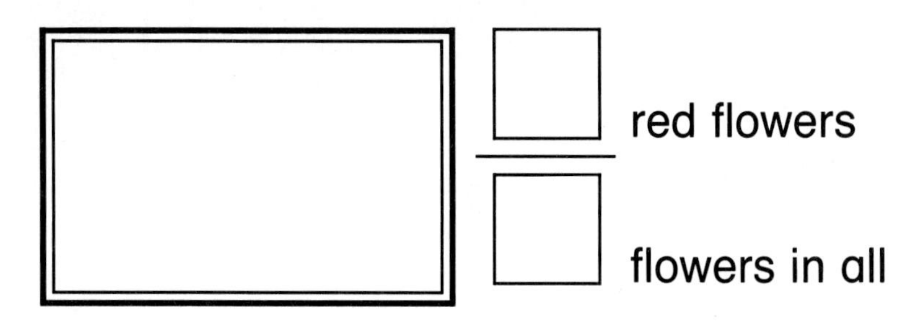

Complete.

2.

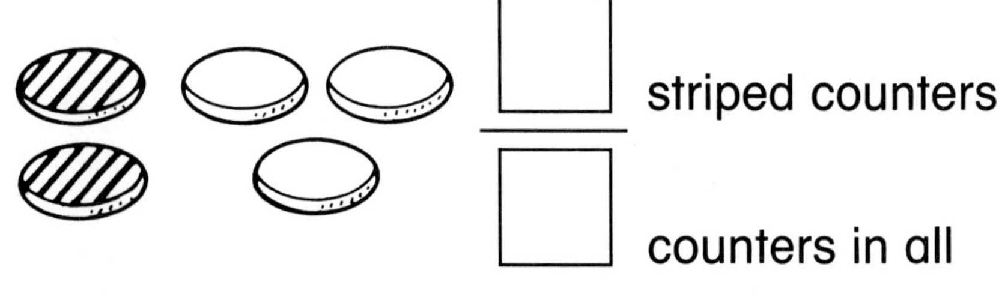

3.

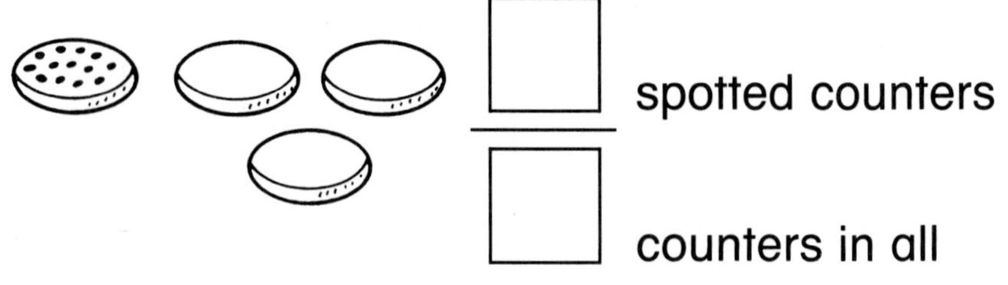

Use before pages 341–342.

Name ______________________

What do you think will happen?
Ring your guess.

1. Will Sam feed his horse carrots?

sure to happen

may happen

impossible

2. Will Karen ride on her goldfish?

sure to happen

may happen

impossible

Write the fraction.

3. What part is spotted?

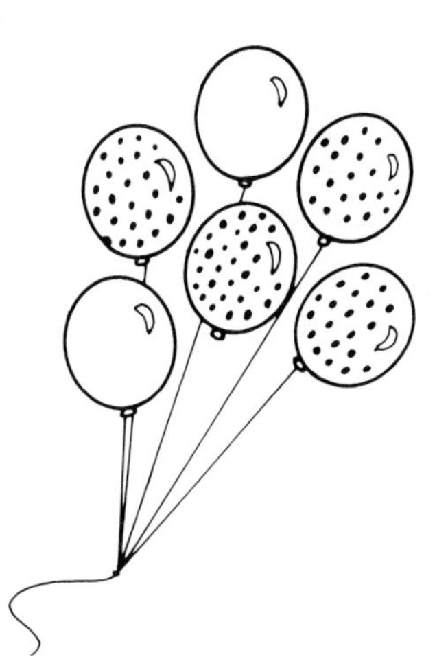

spotted balloons

balloons in all

 Use before pages 343–344.

Name

What do you think will happen?
Ring your guess.

1. If the temperature is 100 degrees, it will snow.

sure to happen

may happen

impossible

Draw a picture. Then solve.

2. There are 3 empty glasses and 2 glasses filled with juice. What part of the group is empty glasses?

Use before pages 345–346.

Name

Ring your guess.

1. If you drop a nickel, on which side is it more likely to land?

   Heads is more likely.

   Tails is more likely.

   They are equally likely.

What do you think will happen?
Ring your guess.

2. Will Joan eat cereal for dinner?

   sure to happen

   may happen

   impossible

3. Will the sun set tonight?

   sure to happen

   may happen

   impossible

  Use before pages 347–348.

Name ____________________

Use the table below to help you choose a class pet.

Class Pet Choices	
Fish	|||
Guinea Pig	𝍸
Turtle	𝍸 |||

1. Which pet do you think the second grade class should get? ____________________

2. Why? ____________________
____

Ring your guess.

3. Will Jerry's plant grow if he waters it?
   - sure to happen
   - may happen
   - impossible

Use before pages 359–360.

Name

How many sides and corners?

1.

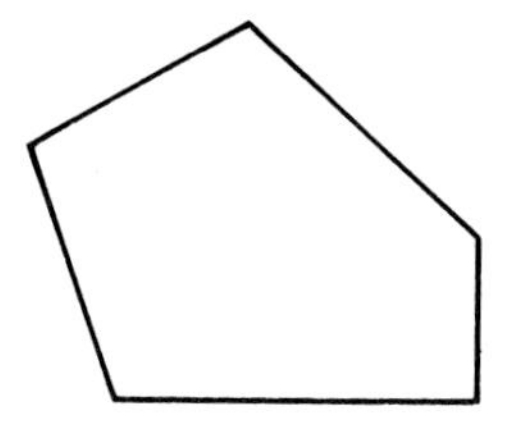

____ sides

____ corners

2.

____ sides

____ corners

3. If you drop a half dollar, on which side is it more likely to land? Ring your guess.

Heads is more likely.

Tails is more likely.

They are equally likely.

Name

Copy the shape.
Write how many sides and corners.

1.

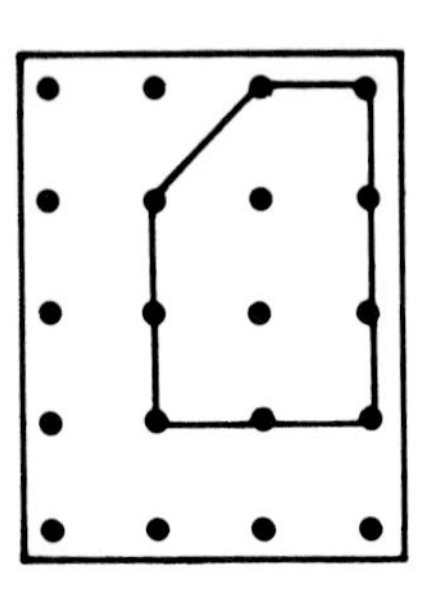

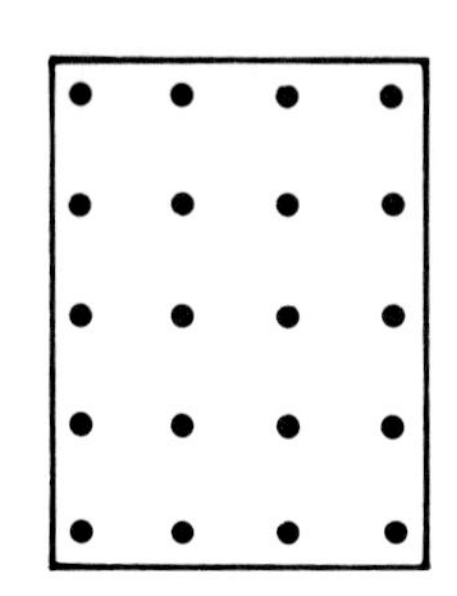

_____ sides

_____ corners

Use the table below to
help you choose a snack.

Snack Choices	
Popcorn	卌 卌 |
Pretzels	|||
Fruit Shakes	卌 |

2. Which snack do you think the second grade class should make? __________

3. Why? __________

Name

Use an inch ruler.
Measure to find the length of each side. Add to find the distance around the shape.

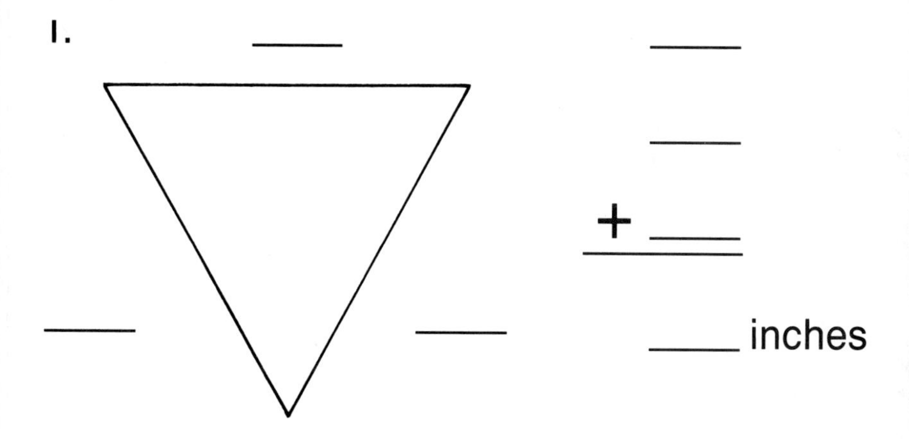

Color the one that is the same shape and size.

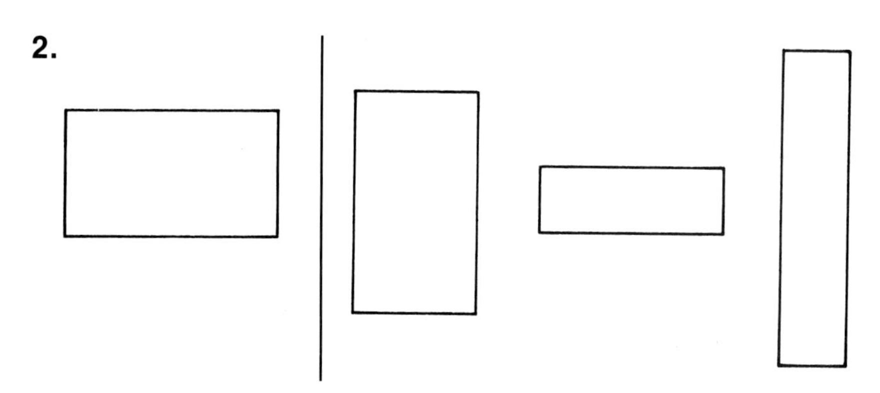

Name

Will the parts match when folded on the line? Ring yes or no.

1\.

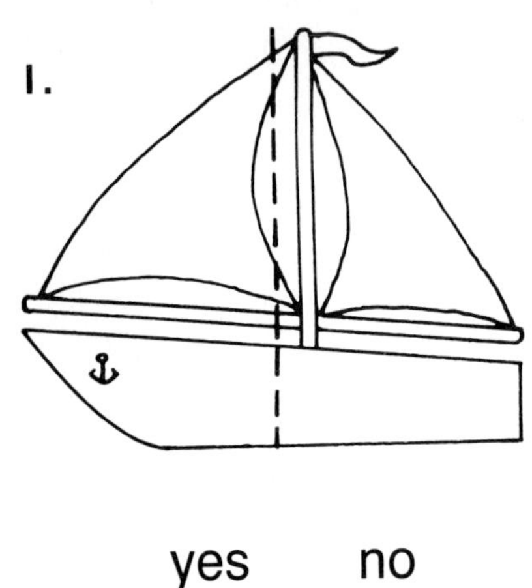

yes no

2\.

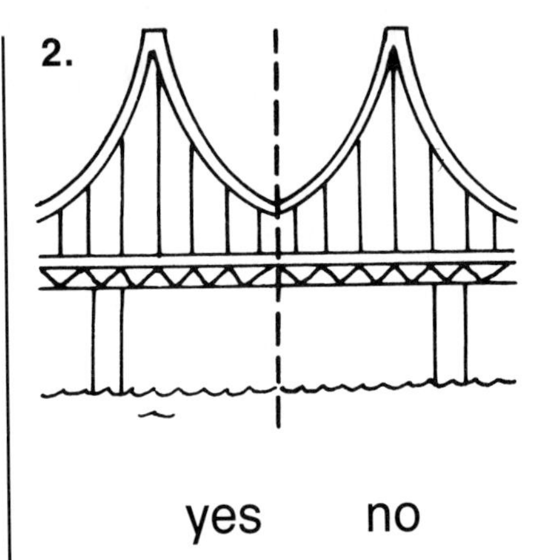

yes no

Write how many sides and corners.

3\.

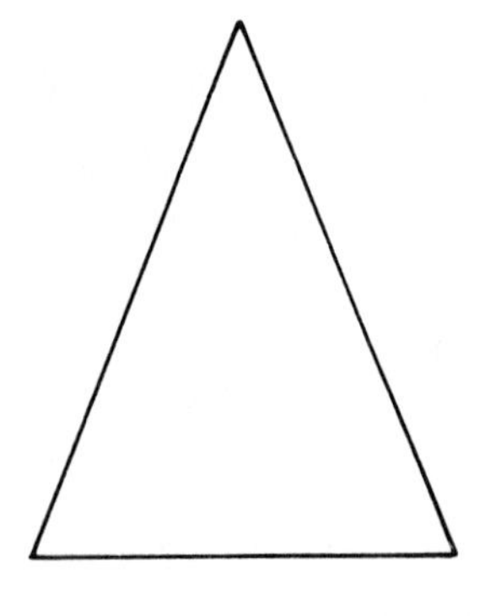

____ sides

____ corners

4\.

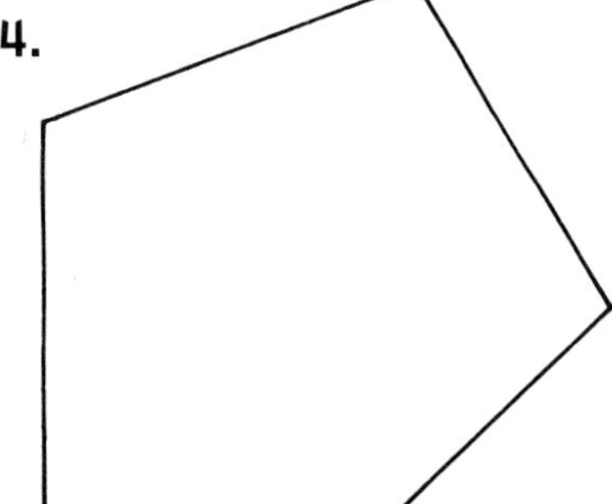

____ sides

____ corners

Name

What will you see if you trace the shape?

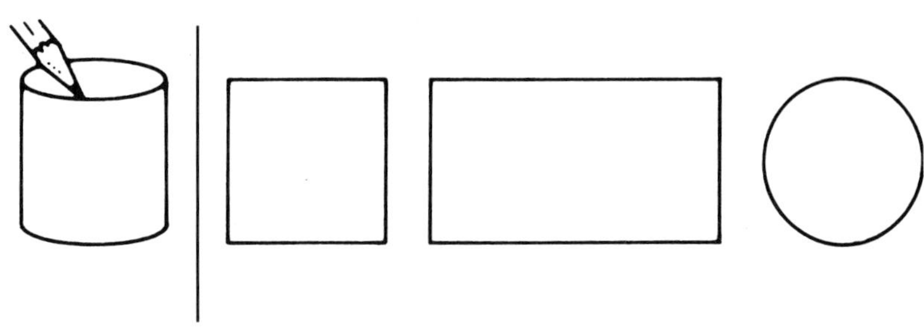

Use an inch ruler.
Measure to find the length of each side.
Add to find the distance around the shape.

2.

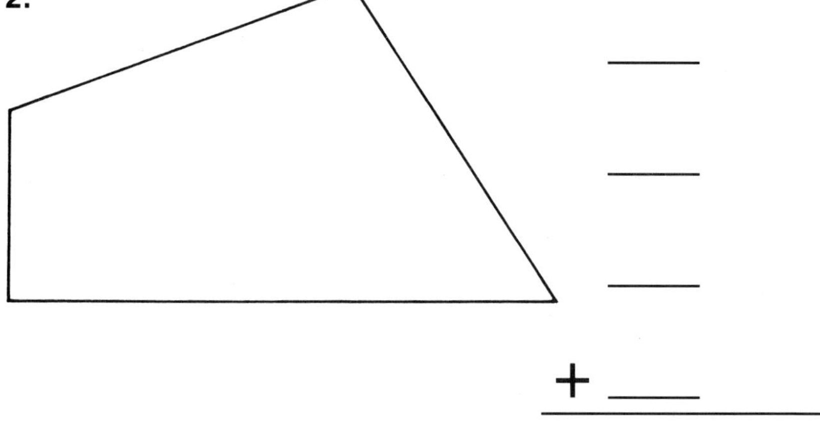

____ inches

Name

Daily Review 122

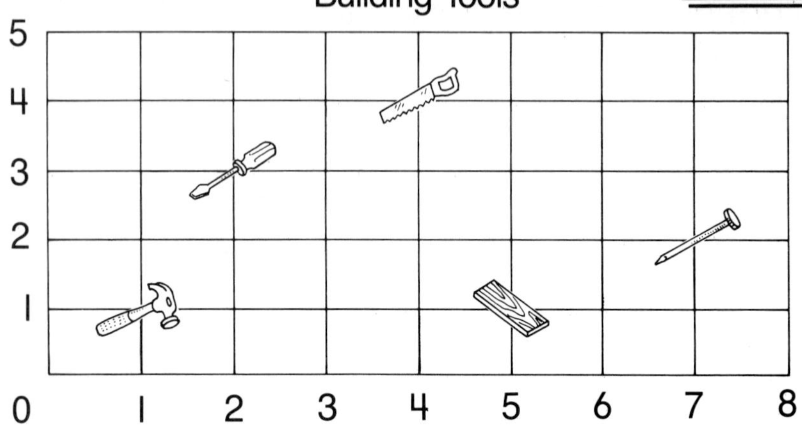

Where is each item on the graph?
Start at 0. Go across. Then up.

1.

item	across	up

2.

item	across	up

Will the parts match when folded on the line?
Ring your answer.

3.

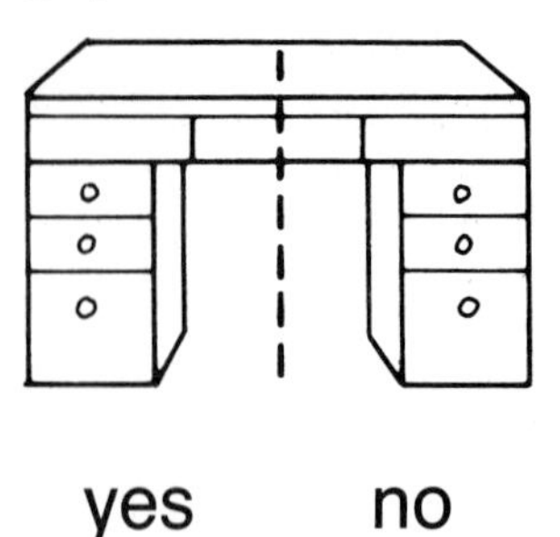

yes no

4.

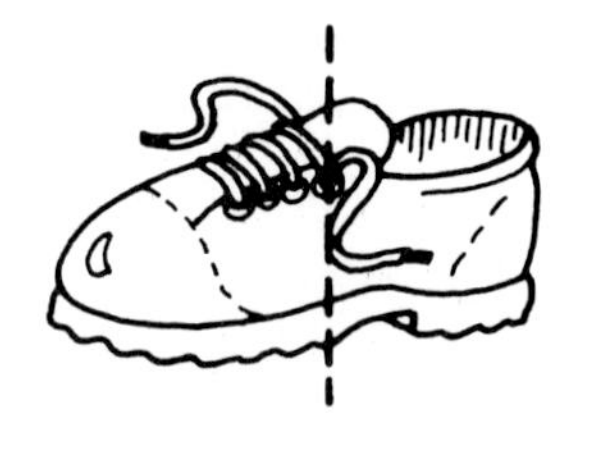

yes no

Use before pages 375–376.

Name

About how much does each container hold?
Ring the better estimate.

1.

less than 1 quart

more than 1 quart

2.

less than 1 gallon

more than 1 gallon

What will Barney see if he traces the shape?

3.

4.

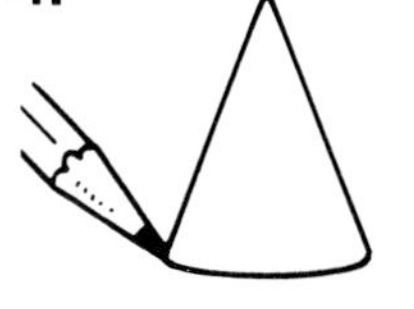

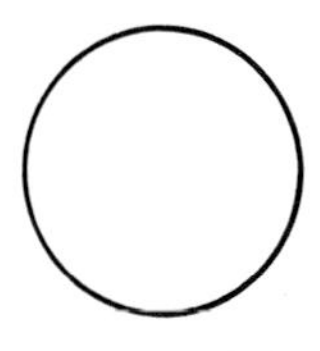

Use before pages 377–378.

Name

About how much does each hold? Ring the better estimate.

1.

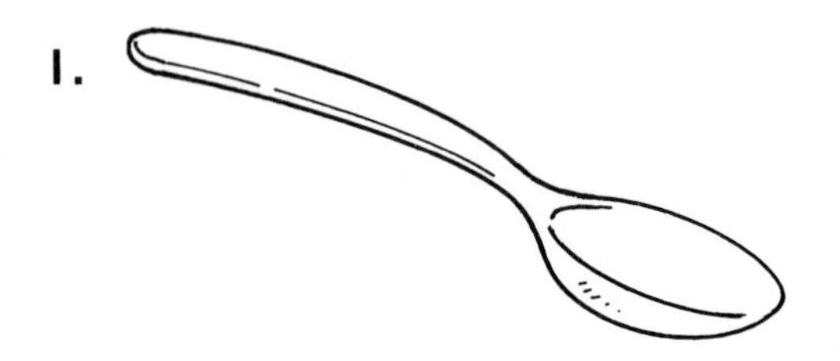

less than 1 liter

more than 1 liter

2.

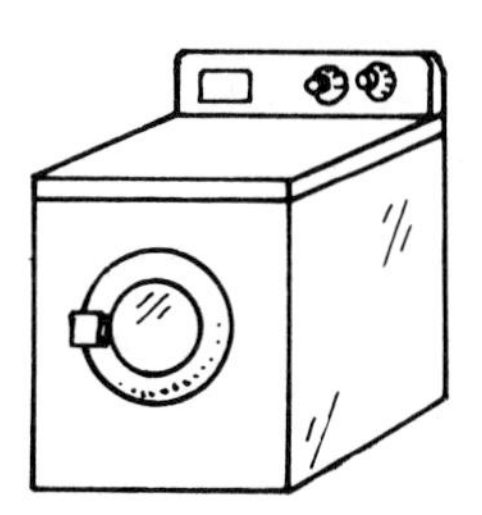

less than 1 liter

more than 1 liter

Will the parts match when folded on the line? Ring your answer.

3.

yes

no

4.

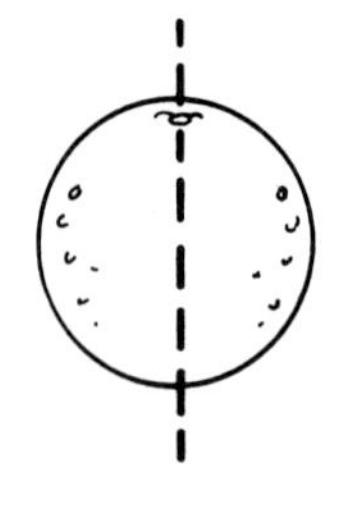

yes

no

Use before pages 379–380.

Name

About how much does the balloon weigh?
Ring the better estimate.

1.

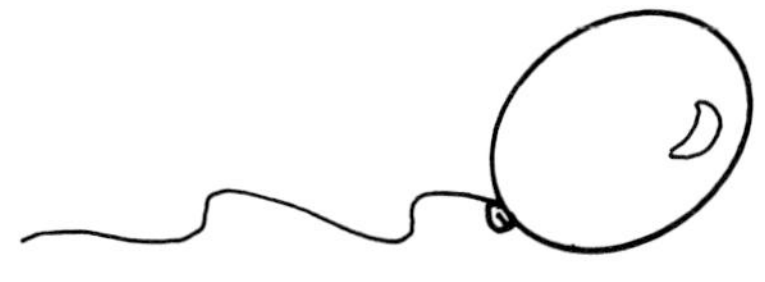

less than 1 pound

more than 1 pound

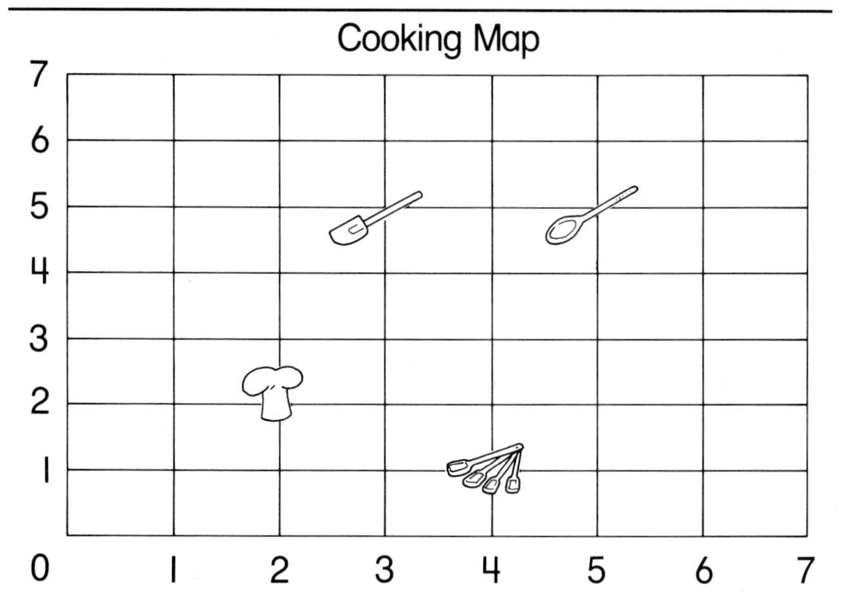

Where is each item on the graph?
Start at 0. Go across and then up.

2.

item	across	up

3.

item	across	up

Use before pages 381–382.

Name

About how heavy is the car?
Ring the better estimate.

1.

lighter than 1 kilogram

heavier than 1 kilogram

About how much do they hold?
Ring the better estimate.

2.

less than 1 gallon

more than 1 gallon

3.

less than 1 quart

more than 1 quart

Name

Write how many degrees.

1\.

______ °F

About how much does each hold? Ring the better estimate.

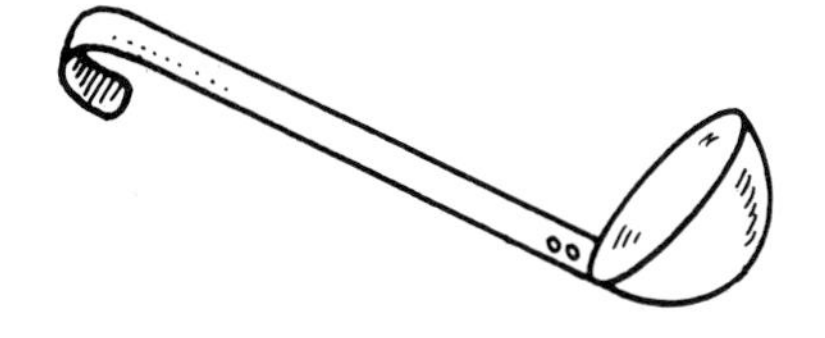

less than 1 liter

more than 1 liter

less than 1 liter

more than 1 liter

Name

Ring the better estimate.

1. What weighs the same as the blocks?

I marble

30 marbles

2. What weighs about the same as your math book?

I quart of milk

25 quarts of milk

About how much does each weigh?
Ring the better estimate.

3. 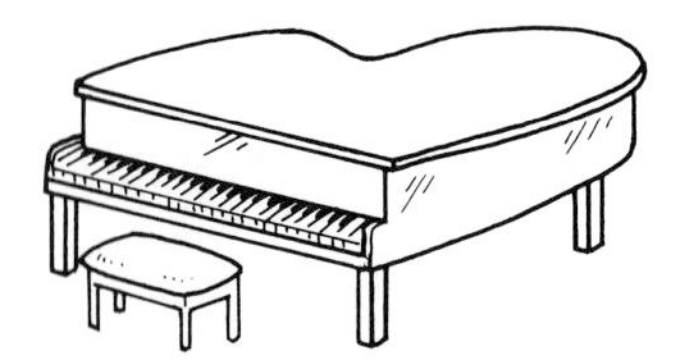

less than I pound

more than I pound

About how heavy is each one?
Ring the best estimate.

4. 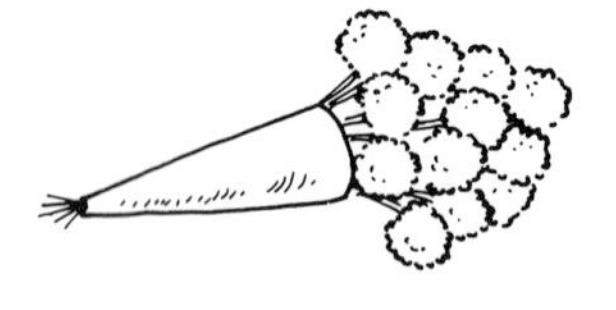

less than I kilogram

more than I kilogram

Name

Ring if you need to trade.
Then add.

1.

$$\begin{array}{r} 452 \\ +\ 37 \\ \hline \end{array} \quad \begin{array}{r} 321 \\ +378 \\ \hline \end{array} \quad \begin{array}{r} 629 \\ +152 \\ \hline \end{array} \quad \begin{array}{r} 313 \\ +\ 44 \\ \hline \end{array}$$

2.

$$\begin{array}{r} 285 \\ +113 \\ \hline \end{array} \quad \begin{array}{r} 425 \\ +\ 61 \\ \hline \end{array} \quad \begin{array}{r} 327 \\ +216 \\ \hline \end{array} \quad \begin{array}{r} 159 \\ +\ 14 \\ \hline \end{array}$$

Write how many degrees.

3.

_____ °F

Name

Ring if you need to trade.
Then add.

1.

$$\begin{array}{r} 343 \\ +128 \\ \hline \end{array} \quad \begin{array}{r} 407 \\ +312 \\ \hline \end{array} \quad \begin{array}{r} 529 \\ +151 \\ \hline \end{array} \quad \begin{array}{r} 375 \\ +\ \ 18 \\ \hline \end{array}$$

Ring the better estimate.

2. What weighs about the same as an apple?

1 pear   8 pears

3. What weighs about the same as a loaf of bread?

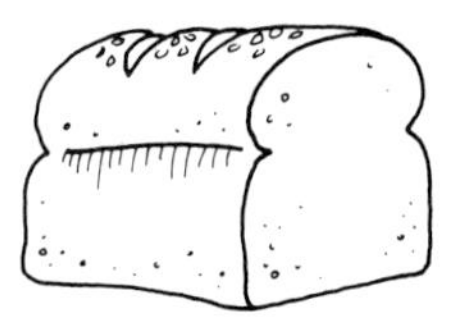

1 carrot   9 carrots

Name

Add.

Trade if necessary.

1.

$$\begin{array}{r} 453 \\ +239 \\ \hline \end{array} \quad \begin{array}{r} 791 \\ +\ \ 14 \\ \hline \end{array} \quad \begin{array}{r} 237 \\ +362 \\ \hline \end{array} \quad \begin{array}{r} 551 \\ +210 \\ \hline \end{array}$$

2.

$$\begin{array}{r} 628 \\ +138 \\ \hline \end{array} \quad \begin{array}{r} 264 \\ +351 \\ \hline \end{array} \quad \begin{array}{r} 453 \\ +426 \\ \hline \end{array} \quad \begin{array}{r} 182 \\ +137 \\ \hline \end{array}$$

Ring if you need to trade.

Then add.

3.

$$\begin{array}{r} 271 \\ +319 \\ \hline \end{array} \quad \begin{array}{r} 142 \\ +139 \\ \hline \end{array} \quad \begin{array}{r} 454 \\ +\ \ 19 \\ \hline \end{array} \quad \begin{array}{r} 229 \\ +120 \\ \hline \end{array}$$

 Use before pages 401–402.

Name ______________________

Use the clues to solve.

1. Judy is looking in her aquarium. She sees fewer than 30 fish. She sees more than 20 fish. There is a 6 in the number. How many fish are in the aquarium? _____ fish

2. Len picks between 60 and 70 apples. The number in the ones place is 2 less than the number in the tens place. How many apples are there? _____ apples

---

Add.
Trade if necessary.

3.

185	520	383	142
+ 372	+ 99	+ 112	+ 56

Name ____________________

Ring if you need to trade.
Then subtract.

1.

274	682	448	961
− 39	−376	−136	−247

2.

453	595	354	509
−121	−238	−127	−207

Ring if you need to trade.
Then add.

3.

537	283	166	224
+ 29	+506	+119	+243

  Use before pages 407–408.

Name

Daily Review 134

Ring if you need to trade.
Then subtract.

1.

366	422	386	659
− 51	− 117	− 128	− 129

Use the clues to solve.

2. Ben is counting the cars on a train. He counts less than 70 cars. He counts more than 60 cars. You say this number when you count by fives. How many cars does Ben count? _____ cars

Name

Subtract.
Trade if necessary.

1.

$$\begin{array}{r} 321 \\ -140 \\ \hline \end{array} \quad \begin{array}{r} 857 \\ -529 \\ \hline \end{array} \quad \begin{array}{r} 562 \\ -292 \\ \hline \end{array} \quad \begin{array}{r} 748 \\ -197 \\ \hline \end{array}$$

Ring if you need to trade.
Then subtract.

2.

$$\begin{array}{r} 438 \\ -129 \\ \hline \end{array} \quad \begin{array}{r} 765 \\ -248 \\ \hline \end{array} \quad \begin{array}{r} 283 \\ -\ \ 37 \\ \hline \end{array} \quad \begin{array}{r} 846 \\ -423 \\ \hline \end{array}$$

Add.
Trade if necessary.

3.

$$\begin{array}{r} 348 \\ +170 \\ \hline \end{array} \quad \begin{array}{r} 673 \\ +129 \\ \hline \end{array} \quad \begin{array}{r} 329 \\ +\ \ 90 \\ \hline \end{array} \quad \begin{array}{r} 172 \\ +116 \\ \hline \end{array}$$

  Use before pages 411–412.

Name

Daily Review 136

Add or subtract.

1. $\$\,4.70 + 3.62 =$ ____ $\$\,6.28 - 2.19 =$ ____ $\$\,5.73 + 2.26 =$ ____

Subtract. Trade if necessary.

2. $648 - 296 =$ ____ $852 - 341 =$ ____ $402 - 181 =$ ____ $625 - 212 =$ ____

Use the clues to solve.

3. Julie has less than 50 seashells. She has more than 40 shells. There is an 8 in the number. How many shells does she have? ______ shells

Name

Continue each pattern.
Add pennies, dimes, or dollars.

1.

$0.35	$0.40	$0.45		

2.

$4.27	$6.27	$8.27		

3.

$1.33	$1.34	$1.35		

Subtract.
Trade if necessary.

4.

$$\begin{array}{r} 457 \\ -266 \\ \hline \end{array} \quad \begin{array}{r} 972 \\ -369 \\ \hline \end{array} \quad \begin{array}{r} 429 \\ -352 \\ \hline \end{array} \quad \begin{array}{r} 641 \\ -330 \\ \hline \end{array}$$

Name ______________________

Daily Review 138

Add or subtract.

1. Yesterday Mr. Lee's students made 23 posters about the school fair. Today they made 18 more posters. How many posters did they make all together?

______ posters

2. Mrs. John's students made 80 popcorn balls to sell at the school fair. They had 14 left. How many popcorn balls did they sell?

______ popcorn balls

Add or subtract.

3.

$2.69	$9.64	$4.30
+ 5.25	− 3.71	− 1.19

Name ______

Daily Review 139

Add or subtract.

1. Kevin read 114 pages in his book last week. This week he read 90 pages. How many pages has Kevin read in his book so far?

______ pages

2. Linda's book has 138 pages. Rita's book has 46 more pages than Linda's. How many pages are in Rita's book?

______ pages

Continue each pattern.
Add pennies, dimes, or dollars.

3.	$2.10	$2.20	$2.30		
4.	$0.18	$2.18	$4.18		

Name ________________

Complete.

1.

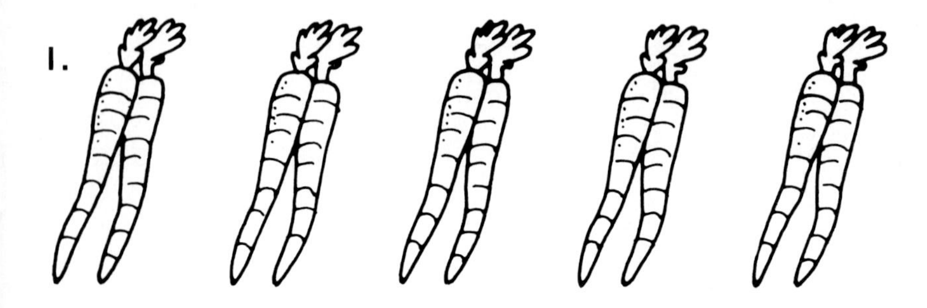

How many bunches of carrots? _____

How many carrots in each bunch? _____

How many carrots in all? _____

Add or subtract.

2. Ron had 72 stamps in his collection. He gave 13 stamps to Carol. How many stamps does he have left?

_____ stamps

Name

Add or multiply.

1.

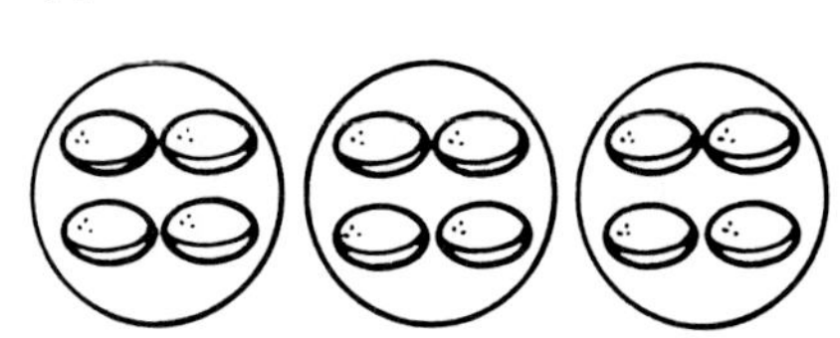

$4 + 4 + 4 =$ ____

$3 \times 4 =$ ____

2.

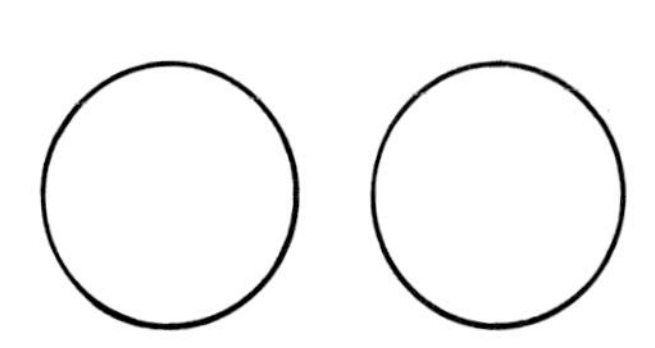

$0 + 0 =$ ____

$2 \times 0 =$ ____

Complete.

How many groups of counters? ____

How many counters in each group? ____

How many counters in all? ____

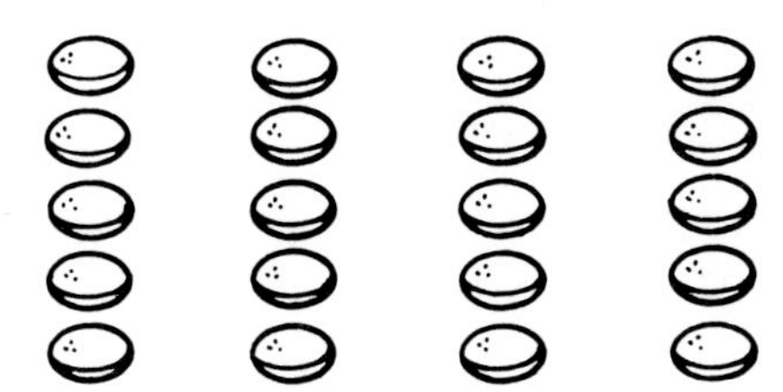

Name

Daily Review 142

Add or multiply.

1.

$3 + 3 + 3 =$ _____

$3 \times 3 =$ _____

2.

$5 + 5 + 5 =$ _____

$3 \times 5 =$ _____

Add or subtract.

3. Yesterday it was 73°. Today it is 9° warmer. What is today's temperature?

_____ °F

Name ______________________

Daily Review 143

How many?

1.

$$\begin{array}{r} 2 \\ \times 7 \\ \hline \end{array}$$

2.

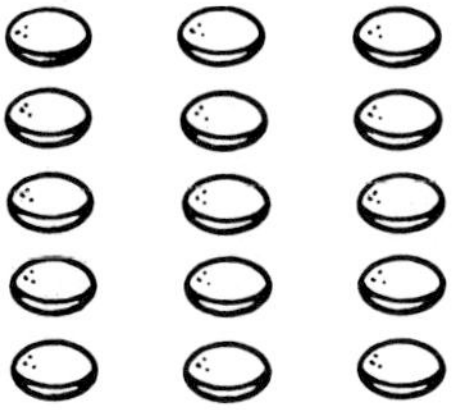

$$\begin{array}{r} 3 \\ \times 5 \\ \hline \end{array}$$

Complete.

3.

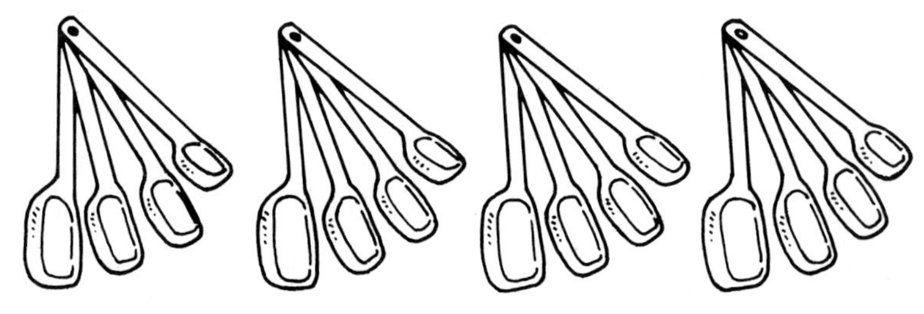

How many groups of spoons? _____

How many spoons in each group? _____

How many spoons in all? _____

Name ______________________

Find the product in two ways.

1.

____
$\times$ ____
____

____
$\times$ ____
____

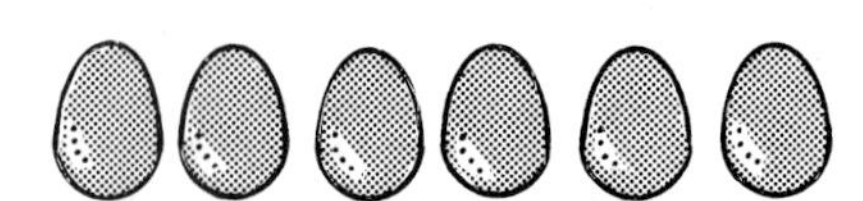

Add to find how many.
Multiply to find how many.

2.

$5 + 5 + 5 =$ ____

$3 \times 5 =$ ____

3.

$6 + 6 =$ ____

$2 \times 6 =$ ____

Name

Kate asked 50 children to name their favorite fruit. This pictograph shows the results.

Favorite Fruit	★ = 5 people
Apple	★ ★ ★ ★ ★
Banana	★ ★
Grapes	★ ★ ★

1. How many people named each fruit?

Apple ______ Banana ______ Grapes ______

How many?

2.

$$\begin{array}{r} 4 \\ \times 2 \\ \hline \end{array}$$

3.

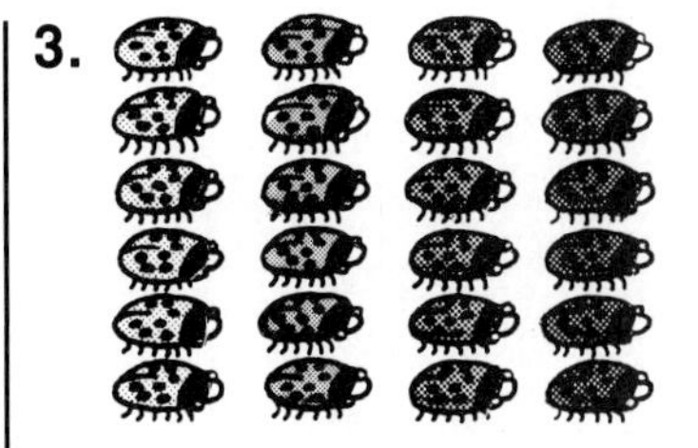

$$\begin{array}{r} 6 \\ \times 4 \\ \hline \end{array}$$

Name

1. How many?

▽ ▽ ▽ ▽ ▽ ▽

▽ ▽ ▽ ▽ ▽ ▽

▽ ▽ ▽ ▽ ▽ ▽

▽ ▽ ▽ ▽ ▽ ▽

▽ ▽ ▽ ▽ ▽ ▽

$5 \times 6 = \_\_\_\_$

Find the product in two ways.

3.

$\_\_\_\_$

$\times \_\_\_\_$

4.

$\_\_\_\_$

$\times \_\_\_\_$

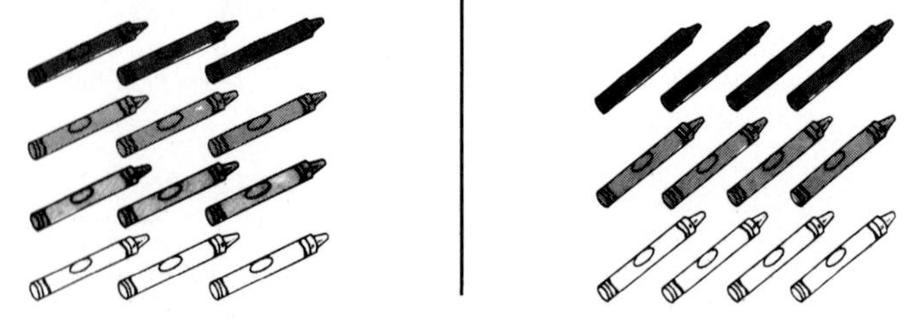

Name

Ring the groups.
Write how many groups.

1.

____ groups of 2 in 12

2.

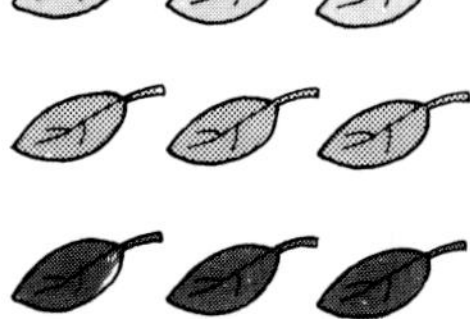

____ groups of 3 in 12

Dave asked 30 children to name their favorite art project. This pictograph shows the results.

Favorite Art Project	★ = 3
Painting	★ ★ ★
Collage	★ ★
Modeling Clay	★ ★ ★ ★ ★

3. How many people named each activity?

Painting ______

Collage ______ Modeling Clay ______